JEWEL OF THE KOOTENAYS

Jewel of the Kootenays
THE EMERALD MINE

Larry Jacobsen

Copyright © Larry Jacobsen 2008. All rights reserved.

No part of this book may reproduced, stored in a retrieval system, or, transmitted by any means without the written permission of the author.

For more information contact:
Gordon Soules Book Publishing Ltd.
1354-B Marine Drive
West Vancouver, BC V7T 1B5
Phone:(604) 922-6588
Fax: (604) 688-5442
E-mail: books@gordonsoules.com

LIBRARY AND ARCHIVES CANADA CATALOGUING IN PUBLICATION

Jacobsen, Larry, 1928-

    Jewel of the Kootenays : the Emerald Mine / Larry Jacobsen.

Includes index.

ISBN 978-0-9781640-1-0

    1. Emerald Mine (Salmo, B.C.)—History. 2. Mines and mineral resources—British Columbia—Salmo Region—History. 3. Salmo Region (B.C.)—History. I. Title.

TN27.B9J33 2008   622'.340971162   C2008-904406-1

Front cover photo: Inspecting cave-in in the Jersey 42 adit. Pictured are Wally Armstrong, Bill Fudicoff, and Ed Lawrence, with John Bishop in the foreground. The photo was taken by Larry Jacobsen in 2007, but the scene could just as easily have been from over 50 years ago.

Author photo: Ed Mansfield

Design and layout by Vancouver Desktop Publishing Centre
Printed in Canada by Ray Hignell Services Ltd.

Unless otherwise attributed, all photos courtesy of the author, Larry Jacobsen.

Acknowledgments

Researching and writing this book has been an intriguing adventure for me, and there are many people I would like to thank, who have contributed to the project in various ways.

First, the book wouldn't have seen the light of day without one person: Ed Lawrence, former mine manager for Canex and present exploration manager for Sultan Minerals. It was Ed who urged me to take on this project in the first place, and he generously shared his resources and stories with me.

I am grateful to the Salmo Arts and Museum Society, in particular to Heather Street, for believing in the project and agreeing to publish it, and for securing funds from the Columbia Basin Trust. I am also grateful for the generous financial assistance received from Sultan Minerals Inc. in Vancouver, as well as for the personal support from Arthur Troup, CEO of Sultan Minerals Inc., including his coming up with the title for the book.

Another person who deserves special mention is Beth Simmons of the Community College of Denver, Colorado, who not only helped me with information about Harold Lakes but also wrote the wonderful foreword.

A large part of the book consists of interviews with over 90 people who either worked or lived at the Emerald mine between the late 1940s and the early 1970s. I am especially thankful to all these people, too numerous to list here, who generously shared their old photographs and personal stories with me. Conducting the interviews has given me the opportunity to meet the special kind of people it took to open and operate a mine, some of them still working long after normal retirement age, not because of financial hardship, but because they are too full of life to quit. Their hospitality and recollections afforded me an appreciation of the mine that I should not otherwise have acquired.

I appreciate the assistance given by the staff at the Chamber of Mines of Eastern BC in Nelson and at the Association of Mineral Exploration in Vancouver, who allowed me to copy material in their possession. I am also thankful for the help and co-operation I received from the museums in Salmo and Nelson for the use of archival material they put at my disposal. I want to thank Nean Allman of Allman & Associates Corporate Communications in Kitchener, Ontario, for her helpful suggestions on publicizing the book. Terry Turner, retired geologist in Riondel, the author of several local histories, generously shared his insights and experiences with me as well.

A number of people lent me material from their personal collections. I am especially indebted to Mel Olson of Maple Ridge for allowing me to copy Canex reports that had been previously published in the *Canadian Mining and Metallurgical Bulletins* from the early to mid-1950s and for permission to reprint them. Earl McLean furnished me with material from a 1976 issue of *Placer News,* and Dennis Hartland lent me a 1976 issue of the *Western Miner* magazine. I am also indebted to the people at the *Northern Miner* magazine for permission to use several articles they had previously published. I want to thank Matthew Allen for allowing me to reprint the article he wrote for *Placer News*, which I am using as an epilogue. It is a fitting tribute with which to close the stories of the mine.

Government House in Victoria kindly provided an image of Charles Arthur Banks, a co-founder of Placer and the Lieutenant-Governor of BC from 1946 to 1950.

I would also like to mention the Salmo high school who ran a contest for a title for this book. We did not use any of the entries, but the best one was submitted by Amanda Shellenberg, followed by Lydia Henderson who placed second.

I am indebted to my editor, Helmi Braches, for helping me to make sense of all the material and to improve its readability, and to Patty Osborne of Vancouver Desktop Publishing for turning it all into a beautiful book. Of course, I accept full responsibility for any errors.

Last, but far from least, I want to thank my wife Linda for putting up with my absences (even when I was in the house) during the past year and a half.

—*Larry Jacobsen,*
Port Coquitlam, August 2008

Contents

The First Fifty Years / 15

The Canex Years / 28

"Being There" / 61

The Dark Side of Mining / 252

Epilogue / 283

Glossary / 285

Appendix 1 / 291

Appendix 2 / 296

Appendix 3 / 319

Appendix 4 / 322

Index / 334

Foreword

Providing materials for civilization, the dangerous occupation called mining spans the globe, crosses international borders, supports entire regions, and creates tight-knit communities. Using powerful oral interviews that supported a complete company record, Jacobsen's history of the Emerald Mine poignantly demonstrates how one small place on the face of the planet can affect people around the world. The oral histories gathered in this immense research project demonstrate the accuracy of collective human memory. Details varied little from one storyteller to the next as the ambitious interviewer travelled or telephoned around the country or across the world to record the tales.

Arriving at an international rendezvous high in the mountains of British Columbia, immigrant miners, mining personnel who had worked overseas, and their wives and families roosted, perhaps briefly, in the company town or in the nearby village of Salmo. A collective force, not one person or group, created a close-knit community, yet one that families entered and left frequently. Mining families recognize that theirs is a "moving" occupation; some miners and engineers may work in as many as fifty mines throughout their lives and live in as many places. Miners know that eventually a mine will play out, the job will end, and they will have to move on. This history demonstrates the fleeting times that miners spent in various locales, mostly in British Columbia. At some point in time, all passed through the famed Emerald Mine.

The Emerald Mine community teemed with an encompassing spirit. Like the families who lived in the Colorado company towns of Climax or Gilman during the 1950s, danger and isolation seemed to create a cooperative camaraderie. Fortunately, the operators of Canadian Exploration Limited understood such sociology and tried to treat everyone compassionately. They hired the wives, widows, and older students, so the miners and workers would stay long enough to see the project through the next day or next phase.

Many of the miners' tales focussed not so much on what happened underground, but on survival in a forested wilderness replete with bears that raided garbage cans daily, snow that buried the town up to fourteen feet deep, and temperatures that dropped to 30° below zero regularly during the winters. The glint in the miners' eyes must have sparkled as they relived their experiences at Canex, telling of accidents or inane events (usually caused by alcohol abuse).

Their wives related primitive living conditions where snow blew through the cracks in the pre-fab houses and forced them to solve problems simply to keep warm, cook, and do the laundry. Their stories amazingly ended by saying they would not have changed the time they spent at the Emerald. Children raised in the scenic, emerald-green paradise of a playground are loyal to their childhood home where families of six shared "duplexes" 20 feet wide by 24 feet long!

Miners, the world's most determined, dogged people, take great pride in their work and welcome the challenge of producing more and more materials under the constant threat of injury, death, and ever increasing environmental regulations. In "Jewel of the Kootenays - the Emerald Mine," Jacobsen captures the essence of that formidable force, "mining mentality."

—*Beth Simmons, Colorado mining historian, Lakewood, CO, USA*

Introduction

It came to me as a shock. In 2006 I was visiting my old friend Denis Hartland at Williams Lake, when he told me that the Emerald Mine near Salmo in south-eastern British Columbia might reopen. The old girl might not be dead quite yet; they had discovered a lot more ore. Since I worked in this mine 55 years ago, I had a personal connection to it.

The name "Emerald" refers to a mine that never produced emeralds. After Canadian Exploration Limited acquired the property in 1947, it became more generally known as "Canex" and came to include several ore bodies in close proximity to each other on Iron Mountain. The Emerald, though the first ore body mined, was just one of six—Emerald Lead-Zinc, Jersey Lead-Zinc, Emerald Tungsten, Dodger, Invincible, and Feeney, the last three being chiefly tungsten with small amounts of molybdenum. Still, the name "Emerald" is often used for the entire mining operation on Iron Mountain.

The Emerald owed its existence and most of its success especially to a handful of men who had the foresight, courage, and special talents needed to make this mine the success it became.

John Waldbeser staked the Emerald claim in 1896 and shipped the first ore nine years later. When Iron Mountain Mines Ltd. purchased the mine from him, he stayed on as the manager, and he was still in that position in 1942, when the Wartime Metals Corporation (Canadian government) expropriated it.

Another significant contributor was Harold Lakes, whose detailed mapping and exploration during the late 1930s and early 1940s led to the discovery of tungsten and the other ore bodies. His maps were so thorough that some of them are still being used, says Ed Lawrence, who is now overseeing the present drilling program.

There was also Charles A. Banks, cofounder of Placer Development Company in 1926 and Lieutenant-Governor of BC when Placer acquired the property from the Wartime Metals Corporation in 1947. He mentored some of the key people who turned the mine into a major success.

These same people later opened a number of other mines in BC that included the Craigmont copper mine at Merritt, the Endako molybdenum mine at Fraser Lake, and the Gibraltar copper mine near Williams Lake, as well as mines in the Philippines and New Guinea.

One characteristic of mining is that what is uneconomical ore today may be profitable later as the technology advances or the commodity prices improve. During the early 1900s, the miners did not bother with ore that did not run from 35% to 40% lead. Fifty years later Canex mined eight million tons of ore that averaged 2% lead and 3.8% zinc.

The Emerald mine operated during two extended periods: 1905 to 1925 and from 1947 to 1973. It appears that it might now be on the verge of yet another life; recent diamond drilling results suggest that there are large tonnages of molybdenum and tungsten ore remaining.

Canex was for a few years the largest producer of tungsten in Canada and the second largest in North America. During World War II and the Korean War, tungsten from China (the largest supplier of the metal) was cut off, and a frantic search was launched for western sources. The metal is used in many steel alloys and, when combined with carbide, is used in a large number of cutting tools.

Canex was the first mining company in Canada to go "trackless." Instead of laying track (rail) for transporting its ore to surface, it pioneered the use of large diesel equipment underground to load and haul the ore. This was not as simple as it sounds, because it created formidable ventilation demands as well as unforeseen opportunities. When the expanded output overwhelmed the capacity of the aerial tramway that carried the ore down to the concentrator, the company was forced to rethink its entire materials handling strategy. This resulted in their installing a system of tunnels and belt conveyors that radically cut those costs.

The mine was the only one known to have a heated Olympic-sized swimming pool for its employees, built with company-supplied materials and volunteer labour. The pool was fed with the cooling water from the giant air compressors used to power the underground drills.

Mining is a risky business, and the odds stacked against success are formidable. Exploring for metals is costly. Finding economic quantities of any is unlikely, and running the gauntlet of today's regulatory agencies is frustrating and daunting. It presently takes over ten years to bring a good mining prospect into production, and the odds of completing it successfully are low. A mining company must not only contend with existing laws, but with the unseen demands of future ones as well. What kind of people will invest their efforts and money on such poor chances of success? A prospector has to be a dreamer, for how else can one account for the fact that people do continue to search for and occasionally open new mines?

Part of the history of the Emerald is the story of the people who worked and lived there. I was fortunate to be able to track down and interview many people between the ages of 54 and 90, who had worked at the mine, as well as the wives and children who had lived there. What struck me about the older people was that many were still active, some of them still working. Not knowing when to quit is one characteristic that set them apart from ordinary folks and is also the quality that made them leaders. One such man, who is eighty, together with his young (67-year-old) partner recently travelled to Colombia in South America to help a mining company modernize its operations. Robert Fulford recently wrote in the *National Post*, "Never retire. Don't even think of it. It's a killer."[1] That attitude describes a number of mining men in this book.

It seems to me that so much of the popular literature casts company towns in a poor light, but the most repeated phrase I heard from the miners' children who grew up at the Emerald was, "It was wonderful!" They would then elaborate on the joy and freedom of roaming the mountainside, swimming, skating, fishing, building forts and tree houses, or playing "kick the can."

The Emerald had more than its share of the special kinds of men and women it takes to make a mine successful. There were many intriguing people I could not interview because they have long since died. Two such people were Russians (one a prince) who had fled with their families during the revolution of 1917. The best I could do was to read their obituaries and talk to those who had known them. Another was an Australian geologist who was so polite that he'd apologize to his chair if he accidentally knocked it over. Yet he was a decorated pilot who had flown unarmed supply planes from India to China over Japanese-held territory during the Second World War. When Canex hired him, they thought they were hiring his father, who was a noted geologist.

Large companies are often perceived as heartless. Yet the management at Canex was able to successfully infuse its workforce with a fierce loyalty. One reason was that the company repaid that loyalty in kind. One man, who lost an arm on the job, was sent to welding school. Later, when he worked at the company's Gibraltar mine near Williams Lake, Canex provided him with a second mortgage for the house he

---

1 Robert Fulford; Printed with permission of Fulford and the *National Post*, Dec. 15, 2007.

built there. When he died at an early age, the company, without being asked to, forgave that mortgage. In another instance when a woman with several young children suddenly lost her husband to illness, the chief accountant urged her to take secretarial training, after which he hired her. Mine management charged her for neither rent nor utilities for almost a year while she was in school.

The Emerald was an important part of the local mining scene. During its last lifespan, it had more employees than the other two nearby mines combined. Although many of these people lived in the townsite at the mine, the impact on the local economies of Salmo, Nelson, and Trail was considerable.

The mine closed in 1973 when the new NDP government changed the economics of mining with a proposed super-royalty. This made it uneconomic to develop the remaining low-grade ore, and the mine was shut down. The property was later acquired by Bob Bourdon and Lloyd Addie, who in turn sold it to Sultan Minerals Inc. This company has discovered important new ore bodies and extensions to the old ones.

Will the Emerald, like the phoenix of myth, arise out of its ashes and return to life? We don't know, but you can bet that the people of Salmo and the surrounding areas will be eagerly waiting for it to happen.

ONE
The First Fifty Years
1896 TO 1946

The second half of the nineteenth century was pivotal for what was to become British Columbia. During this era, first Vancouver Island, and then the rest of the province became Crown colonies of Great Britain. They shortly merged and became known as British Columbia, with James Douglas as its first governor in 1864.

The first Caucasians to enter British Columbia in large numbers were the gold miners. Thirty thousand of them poured into the Fraser Canyon in 1858 and prospected the river from Hope to Lillooet until the gold petered out in 1860. This was followed by a find at the mouth of the Pend Oreille River in 1859, the 1860 gold strike at Rock Creek, and then the 1862 Cariboo gold rush at Barkerville. From then on gold was found in other places as well: Tulameen in 1886 and Rock Creek again in 1887. Meanwhile silver, lead, and zinc were discovered in the Kootenay Lake area, with the Bluebell claims being staked in 1882 and the Copper Mountain copper ore body near Princeton discovered in 1888. The first important find of ore in the Nelson area was the Silver King deposit, ten kilometres south of Nelson.

Access to the mineral finds was usually difficult, but at least with gold at twenty dollars per ounce, shipping it to market did not consume most of its value like that of the other ores. With silver-lead ores, the freight amounted to an important portion of their total value. Proximity to navigable water helped, but it was not until rail transport became available that mining for the other metals really took off. This was certainly the case with the Salmo area.

Another problem was price volatility. Whereas gold remained steady at twenty dollars per ounce, silver and lead fluctuated greatly. In 1895, silver sold for 6½ cents per ounce, whereas it had been $1.09 in

1890. Prices for base metals varied equally. Even with good prices, these ores had to be high-grade to make them economical to mine.

There were fifteen placer claims on the Salmon and six on the Pend Oreille rivers in 1893, but when the Nelson and Fort Sheppard Railway arrived in 1894, prospecting in the area zoomed. The first of the Sheep Creek mines was staked in 1896, although serious production was not begun until 1902. This camp, which came to include such names as the Mother Lode, Nugget, and Reno in one group, the Queen and Sheep Creek in another, as well as the Kootenay Bell and the Gold Belt mines, ranked sixth in amount of lode gold produced in BC. From its beginning until 1951, this camp produced 736,000 ounces of gold, 365,000 ounces of silver, 377,000 pounds of lead, and 312,000 pounds of zinc.[2]

Other mines in the area that began production after the railway came were the Yankee Girl and the Ymir Gold Mine, as well as the Arlington and Second Relief near Erie.

New railroads proliferated in Western Canada from 1880 to 1900, with the Canadian Pacific Railway (CPR) reaching Port Moody in 1886, followed by a line to Nelson a few years later. The railroads had a huge impact on mining because they made it possible to transport people and heavy supplies to the mines, haul the ore to the smelters, and ship the metals to markets from areas that had hitherto been accessible only by water or pack trains.

The spread of the railways brought not only miners to the areas they served, but also sawmills, logging operations, and farming, followed by schools, churches, and retail stores. The railways made it possible to mine ores that would not otherwise have been of economic value, and one such property was the Emerald Mine near Salmo. Whereas the first mines in the Salmo area produced mostly gold, the Emerald had high-grade lead ore with a couple of ounces of silver per ton in it as well.

There was another event that would later impact mining in the West Kootenays. A geologist named Arthur Lakes came to America from Cornwall, England, and settled at Golden, Colorado. In addition to becoming a noted dinosaur fossil hunter, he was also regarded as a de facto co-founder of the Colorado School of Mines, and he was its first professor of geology. He spent some time prospecting in southern BC and wrote papers on the mining potential of a number of mineralized areas. Lakes died at Nelson, BC, in 1917 at age 73 and was buried in the Nelson cemetery. More important from our perspective is that his two sons, Arthur Jr. and Harold, would both become geologists and

Arthur Lakes Sr. Born in Cornwall, England in 1844. Died in Nelson, BC in 1917. Undated photograph. Courtesy of Beth Simmons of Colorado.

2 BC Minister of Mines, *Bulletin 31*.

spend much of their lives in the West Kootenays. The brothers were both involved in a number of mines in the Salmo-Nelson area, including the Emerald Mine. Arthur eventually moved to Nevada, but Harold stayed and was to become the manager at several Salmo area mines, including the Reeves McDonald, the HB (named after the prospectors who staked it: P.F. Horton and H.M. Billings), and finally the Emerald. It was his work at the Emerald that would become pivotal to that mine's success.

■ ■ ■

Rollie Mifflin came to Salmo in April of 1900 at the age of thirteen. In his book about Salmo he writes that the village of Salmo was first called Salmon Siding but was renamed Salmo, which is the scientific name for the species of fish to which both salmon and trout belong. The name of the community was probably changed in order to differentiate it from the many other places containing the word "salmon," and it may also have made sense to the postal authorities to shorten the name.

Mifflin states that in 1900 the Salmo population consisted of 46 men, 12 women, and 17 children. Apparently there had been nine hotels and seven saloons in Salmo in 1897, but when Rollie arrived only three saloons were left.

Hand-mucking in a narrow drift. 1940s. Note the men are wearing carbide lamps on their hard hats. Courtesy of Joel Ackert.

Mifflin's stepfather, John Waldbeser, set up and ran a sawmill near Waneta in 1892. There, according to Rollie, he sawed lumber for a placer mine. When the mining boom hit Rossland in 1893, Waldbeser partnered with a chap named Louie Blue in another sawmill venture. Waldbeser managed the mill and Blue managed the money. When the partnership broke up, the partner had all the money.

Waldbeser sometimes prospected together with Bob Reeves, whom he had met at Waneta in 1892. On a trip up the Pend Oreille River they found ore and staked the property that was to become the Reeves McDonald mine. In 1896, accompanied by a man named Peterson, Waldbeser staked the Emerald. He was also one of the original owners of the Hideaway mine, and he and William McArthur together were the first to ship ore from the Salmo area. From 1896 to 1898 Waldbeser was a sawyer for Lavin Lumber at Salmo.

In 1905 Waldbeser, with his stepson Rollie Mifflin, then aged eighteen, pioneered the shipment of ore from the Emerald and the nearby Jersey claims. The operation began with sorting the ore, placing the high-grade into burlap sacks, tying the sacks, and placing them in a raw animal hide with the skin side out. The edges of the hide were then sewn together, and each load was dragged on the snow by a horse down the trail to the valley below. In February Rollie, using two horses, began rawhiding ore down the mountain from the Emerald mine. It was tricky work, for the trail dropped 2,000 feet in 2½ miles, an average grade of more than 15%. Sometimes the load would leave the trail and

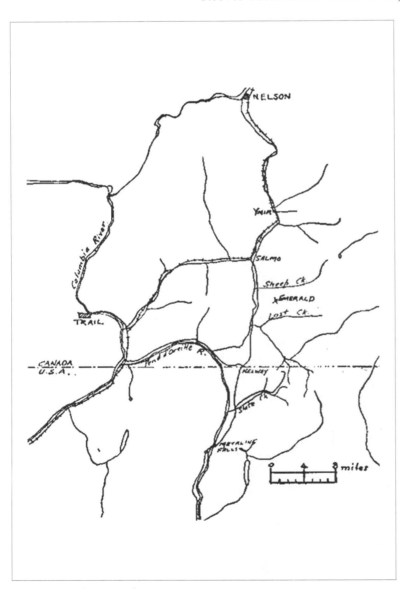

Map showing location of property.

Rawhiding from the Silver King mine, 1888 (?), West Kootenay Mountains, near Nelson BC. The earliest transport from the mine was by rawhiding down a 4.5-mile trail. The mineral-bearing rocks were separated, bagged in raw animal hides, skin side out, the edges sewn together. The packhorses dragged the bags to the trailhead and down the steep mountain trails in winter. Sometimes they wore metal cleats to keep them from slipping, and at times sat and slid in slippery conditions. The lead horse was named Rawhide. Courtesy of the Nelson Museum Archives.

spill into the trees, from where the heavy sacks of ore had to be carried back to the trail and reloaded onto the rawhide before continuing down the mountainside. Even with these difficulties it was relatively easy to pull the load down the hill, but going back up was a different matter, for the rawhides were then loaded with supplies for the mine.

From the bottom of the mountain the ore was hauled by sleigh (or in summer by wagon) to Salmo, from where it was loaded into rail cars and shipped to the smelter at Trail. The ore had to be rich to make it worthwhile to get it from the mine to the smelter.[3]

The first mention of Iron Mountain in the *Minister of Mines Annual Reports for British Columbia* is found in the report for the year 1895:

> Owing to transportation facilities offered by the Nelson & Fort Sheppard Railway, a great deal of attention has been paid by prospectors to the district lying immediately south of Nelson and extending for a distance of 35 miles. Over 200 claims have been recorded, which are tributary to that line. The camp which has come mostly into prominence is that section between Lost Creek and Sheep Creek, which empty into the Salmon River, and known as Iron Mountain, being about 10 miles southeast of Salmon Siding. Experts claim that the ore from this locality is of the same quality as that produced by the War Eagle and Le Roi

---

3 The foregoing is based in part on material from the book, *The Early Salmo Story and Other True Stories* by R. W. Mifflin, published in 1971 and sponsored by the Salmo District '71 Centennial committee.

Ymir in the 1940s. Courtesy of the Nelson Museum Archives.

Ymir early days. Courtesy of the Nelson Museum Archives.

mines on Red Mountain. Assays from the surface have shown as high as $70 in gold per ton. Up to the present, however, little has been done to prove the permanence of the veins.

The Nelson Mining District was a busy area in 1907, with a number of mines (mostly gold) operating in the Salmo-Ymir area. The Minister of Mines report for the year makes its first mention of the Emerald Mine on Iron Mountain as well as such names as the Queen, Kootenay Bell, Mother Lode, and the Nugget in the Sheep Creek camp. At Ymir was the Yankee Girl and at Erie, the Arlington. Other mines in the area were the Eureka, the Queen Victoria, and the Granite-Poorman. Except for the Emerald (lead-zinc) and the Queen Victoria (copper), the rest were all gold mines with minor amounts of lead, zinc, and silver content.

John Waldbeser operated the Emerald and the Jersey with a small crew of a half dozen men until 1909, when he sold the property to Iron Mountain Ltd. but stayed on as the manager and as a shareholder. Rollie Mifflin worked with him and became the treasurer of the company. Under the new owners, the mine kept being referred to as the Emerald. From then on ore production increased somewhat; a lot of development work was also done. Miners' wages that year were reported as $4.00 and $3.50 per day, for a 10-hour day.

During the early days of the twentieth century, rock-drilling was usually done manually by "hand-steeling," either singly or by

1919. Ore teams hauled the lead-zinc ore from the Emerald mine into Salmo, from where it was shipped to the Trail smelter by rail. *Courtesy BC Minister of Mines Annual Reports.*

JEWEL OF THE KOOTENAYS / 21

Table 1.1 Reported Production from the Emerald Mine.

Year	Reported ore shipped	Concentrates shipped	Lead (lbs)	Net smelter receipts	Crew size	Comments
1907	560			$10,000		
1908	426			$7,000		No summer shipments due to poor prices.
1909	1,068		764,292	$5,714	6	Iron Mtn. Ltd. purchase mine. Shipped ore averages 37% lead and 1.5 oz. silver per ton.
1910	1,679					
1911	2,000				8	Ore averages 38% lead and 2 oz. silver per ton.
1912	1,560				8	2 to 4 horse teams kept busy hauling ore to Salmo.
1913	1,100				4 – 9	Labour problems— one third production.
1914	1,136		837,428	$24,239		Shut down 3 mo due to war. Costs $4/ton to haul to Salmo plus $7.50/ton for freight and smelting costs.
1915	1,109				10 – 30	12,000 tons to-date.
1916	1,346				20	Built a small sawmill on the property at cost of $1,000.
1917	4,448				45	Crusher erected.
1918	2,092				20 – 30	Air drills used in Emerald—Hand work at Jersey. Lead price $0.08/lb.
1919	219					50,000 tons shipped to date. Concentrator erected. Price fell to $0.05 after Armistice, then recovered to $0.08.
1920		900			15	Lead @ $0.08/lb—dropped to $0.04 by year-end; the same as the prewar price.
1921						No work done this year.
1922	119					No details available.
1923		223				No details available.

The above information is a summary of what was available from the Department of Mines Annual Reports. The smelter had a policy of penalizing ore that carried too much zinc, but the Emerald ore, averaging only 6 to 7% zinc, was not penalized.

two-man teams. Joe Montgomery, whose story appears later on, explained how it was done:

> I learned to hand-steel when I was at the Tulameen. An old prospector borrowed me from my employer for a couple of days. His name was Joe Levesque. He was a brawny guy who didn't wear a shirt nor bother himself about the mosquitoes or things like that. He came over to the mine where I worked and said he wanted to borrow me, and in return he would put in a couple of days for the company later. So we went over and put in a few holes on his claim, and that was his assessment work for the year.
>
> To drill a hole in the rock, one man holds a sharpened drill steel, while the other man hits it with a ten- or twelve-pound sledge hammer. The man holding the steel turns it after each blow. He also pours a little water into the hole to create a mud-slurry from the rock cuttings as the hole progresses. When this slurry begins to get stiff, he yells, "Mud!" He then yanks out the drill steel and scrapes the slurry out with what is called a "spoon." The spoon is a long thin steel rod that has been hammered flat on one end, and this flattened end is curled back on itself to create a scraper.
>
> He then takes his turn on the hammer, while the other guy turns the steel. The man wielding the hammer tends to be careful not to miss, because his turn to hold the steel is coming and he doesn't want his knuckles skinned (or broken) by a careless swing of the hammer.
>
> One element of the business is that I had to learn to sharpen the steel every day. We did this by trial and error. There was a farmer down in the valley who owned a forge. So we got permission to use it. We heated the steel until it was cherry red, hammered it out into the proper shape—it couldn't be too tapered. Then we dunked it in water to harden it. Once

Cliff Rennie, a retired geologist I interviewed, tells the following story about hand-steeling:

A little Irishman was turning the steel for a big Englishman who was swinging the hammer. He missed occasionally, bruising the Irishman's knuckles. Finally the little Irishman had enough and said, "Hey, let me skin for a while."

1930s Hand steel competition. Note the starter steel flying through the air as the fellow competitor inserts a new steel on the fly, without the hammer-wielder changing his tempo. From the George Murray Family, courtesy of Joel Ackert.

Would you like to hold this steel while someone swings at it with a 12-pound hammer? From the George Murray Family, courtesy of Joel Ackert.

all our steel was sharp, we took them up the hill and resumed drilling.

The first steel we tried shattered. It just flew apart. We had made it too hard. The next one bent because it was too soft. We had cooled it too slowly. After a couple more trips to the forge we finally got it right. I later got a book which explained in detail how to go about the process of tempering the steel. When we followed the instructions in the book we usually got the proper hardness.

A hole scraper and a "hand drill" steel at the Don Endersby Museum, 2007.

I've seen guys single-jacking. They'd use a four-pound single jack and a strap from its handle to their wrist to keep from letting go of it. They then hit the steel with the regularity of a machine. They can bash a drill hole up anywhere they want it to go. It's hard work, really hard work. I have to admire them. I've gone into a lot of old mines and could see they have been hand-steeled. You can't help but admire the men who did that.

The Emerald Mine continued to be the major lead producer in the Nelson Mining District, but the nearby HB mine, which was worked by a small crew of leasers, also shipped lead-zinc ore to the smelter in Trail. With the onset of World War I, Iron Mountain had a problem finding miners and closed for the last three months of that year but reopened at the beginning of 1915. The best year was 1917, when 45 men mined and shipped almost 4,500 tons of ore to the smelter in Trail [see Table I.1].

In 1919 the company installed a concentrator, which eventually, after W. DeWitt had remodelled it in 1920, worked well. Nine hundred tons of concentrates were shipped that year.

A single-jack and two double-jack hammers at the Don Endersby Museum, 2007.

The price of lead fell to five cents a pound after the armistice but recovered to eight cents in 1920. There is no record of prices for the next two years, when there was no reported production at Iron Mountain. It seems that the entire mining in the area came to a halt, for in 1923 only the Kootenay

Belle and the Emerald are mentioned in the annual reports from the Ministry of Mines.

The following year Iron Mountain retained Arthur Lakes Jr. to provide it with detailed geological mapping of the property. His brother Harold assisted him, and the mine shipped concentrates to the smelter again for that year and the next.

In 1926 the property was closed, except for some minor development work, and remained closed until 1938, when Harold Lakes was again hired to supervise development work.[4] (His brother Arthur had moved to Vancouver.) During 1939 and 1940 a small crew drove crosscuts and drifts on the Emerald, and on what would become known as the Dodger property. In 1941 Lakes with the help of two men worked for the entire year, doing detailed geological mapping of the entire property between the Jersey and the Dodger mines, an area one and one-half miles long by one-half mile in width. His mapping led to the discovery of altered limestone beds containing both molybdenum and scheelite (tungsten) ore.

Harold Lakes all duded up in 1913. Courtesy of Peggy Green (niece of Barbara Lakes).

■ ■ ■

When the Canadian government learned of the tungsten, it immediately tried to acquire the property through the Wartime Metals Corporation (WMC) because Iron Mountain Ltd. did not have the capacity to develop the mine at a desirable rate. When a deal acceptable to Iron Mountain Ltd. could not be reached, the government instituted expropriation proceedings. Meanwhile, under the direction of Lakes, Iron Mountain had hired more men, built a new dry, and proceeded with diamond drilling, dozer stripping, and trenching until August 17, 1942, when WMC took over.

Iron Mountain had asked $1,750,000 for the property, but the government paid it only US$424,000.

There are two curious events; according to the 1942 BC Minister of Mines report, the WMC took over operation of the property on August 18, 1942, but the news clipping below indicates it did not file expropriation proceedings until November 25, two months later. The word "expropriate" does not appear in a single report. We know of it only because of the MP's questions in the Canadian House of Commons the following year.

The WMC immediately began to build the infrastructure (except for

Harold Lakes at the Emerald Mine, late 1940s. Courtesy of Peggy Green (niece of Barbara Lakes).

---

4 *Nelson Daily News,* Dec. 28, 1956 – Harold Lakes Obituary.

Esling Obtains Further Data on Emerald

OTTAWA, March 12 — Further information on the Wartime Metals Corporation operation of the Emerald Mine, Tungsten property at Salmo was given in the House of Commons recently when W. K. Esling, Member for Kootenay West, presented a list of questions to the House. The questions and answers follow:

Question — Has the Government, through the Wartime Metals Corporation, or any other department or agency, acquired the Emerald Mine at Salmo, BC?

Answer — The Government acquired the Emerald Mine by expropriation proceedings filed on Nov. 25, 1942.

Question — If so, what price was paid for the property?

Answer — The original owner claimed $1,750,000 but after negotiations agreed to accept $424,000 in U. S. funds.

Question — In whose name does the title at present rest?

Answer — The Canadian Government.

Question — Has the original owner been fully paid?

Answer — No. The legal details have not yet been completed.

Question — What sum has been spent in development?

Answer — $829,100.60.

Question — How many tons of concentrate have been produced since the opening of the property?

Answer — High grade concentrates 132 short tons of 71.07 per cent WO/ Low grade concentrates: 267 short tons of 15.07 per cent WO3.

Question — To whom have these concentrates been sold?

Answer — The low grade concentrates have been sold to the Metals Reserve Company. The high grade concentrates are being retained by the Canadian government.

Question — What sum has been realized from their sale?

Answer — Payment by Metals Reserve Company will be based upon the final assay of the concentrates but it is estimated that approximately $70,000 will be received for the low grade concentrates so sold to date to Metals Reserve Company.

Question — Has this tonnage been shipped or is it still stored at the mine?

Answer — The low grade concentrates have been shipped to Metals Reserve Company. The high grade concentrates are stored at the mine.

Question — When did this property cease to operate?

Answer — Mining ceased Sept. 10, 1943. Milling ceased Sept. 21, 1943. Operations closed Oct. 15, 1943.

A concentrator was built here in 1919 to process lead ore. After the mine ceased production in 1925, the mill sat idle and the concentrator was consumed during a forest fire in 1934. The Wartime Metals Corporation built a new tungsten mill on the site in 1943 and operated it for six weeks before shutting it down. Canex reopened the mill in 1947 and operated it for almost two years before switching it over to process lead-zinc ore at 300 tons per day. Over the next twenty years, the company gradually increased the mill capacity to 2,500 tons per day. A new mill was constructed up on the mountainside to process the tungsten ore, when production of that ore was resumed in 1951. Courtesy BC Minister of Mines Annual Reports.

Bucket on the Canex aerial tramway. Late 1940s to early 1950s. Courtesy of Joel Ackert.

The lead-zinc mill beside Highway 6. The first concentrator was built here and began milling lead-zinc ore in 1919. It burned down during a forest fire in 1934. The War Metals Corporation built a new mill on the site in 1942 and operated it for only six weeks. When Canex purchased the property, tungsten was milled there for almost two years before it was converted it to lead-zinc. The mill was expanded several times and processed lead-zinc ore until 1970, when it was shut down. It was removed from the site in 1973/4. Courtesy of Sultan Minerals Inc.

the concentrator, which it contracted to the Consolidated Mining and Smelting company or CM&S) to put the property into production without delay. Under the supervision of E. E. Mason, camp facilities for a crew of 100 were built, a 6,000-foot aerial tramway from the mine down to the mill was constructed, a temporary diesel compressor unit was installed, and a power line to the mine was built. Development work to the end of 1942 consisted of drifts, diamond drilling, and trenching. The initial crew of 50 men quickly expanded to 150 by the end of the year, with CM&S supplying the management and labour.

During 1943 the crew grew to a total of 250 people before being cut back to 160 after the mine went into production. The camp, including three duplex cottages, accommodated 120 men. Included in the construction that year were a blacksmith shop, explosives storage, a primary crushing plant, and a 30-man bunkhouse adjacent to the concentrator. The tramline from the mine to the mill was completed, as was a modern assay office. The concentrator was finished in June and went into operation on August 1st, milling 200 tons of ore per day.

And yet, despite this major effort in bringing the mine up to a modern standard of production, it was closed six weeks later on orders from the WMC. The reason given for the closure was that the need for the tungsten was no longer urgent. No mention was made of any difficulty in producing an acceptable grade of concentrates, but given the problems Canex would have in doing so a few years later, one may wonder if this played a part in the decision to close down the operation so soon after start-up.

TWO
The Canex Years
1947 TO 1973

After the Second World War, when the Canadian government did not have a dire need of tungsten any longer, the Emerald mine went up for sale. This was when a new player came onto the scene: Canadian Exploration Limited, or Canex.

Canex was a wholly owned subsidiary of the gold placer mining company Placer Development Limited. It was incorporated after World War II when the Placer management decided to pursue lode mining too. The new company's first property was the Emerald Mine, which it purchased in 1947 from the government-owned Wartime Metals Corporation for a down payment of $50,000, with a further $900,000 to be paid for out of profits. The Canex down payment was not the highest, but their overall price appears to have been the most attractive to the government. The Consolidated Mining and Smelting Company (CM&S) had also tendered on the property but was unsuccessful. There was some speculation that C. E. Banks, one of Placer's directors, had an inside track because of his connections to the federal government during the war. There may have been hard feelings between the two companies because CM&S had wanted the property rather badly. At any rate, most of the lead-zinc concentrates from the mine were sold to smelters in the United States and it would be a number of years before any of it went to the CM&S smelter at Trail.

Charles Arthur Banks must have been a remarkable personality, and the story of Canex wouldn't be complete without saying a few words about him. Banks was born in New Zealand in 1885 and died in Vancouver in 1961. He was trained in mine management and came to British Columbia with his French-born wife in 1912. He managed the

Jewel Mine near Greenwood in 1915/16, and both he and his wife served overseas during World War I. He was the manager of the British Columbia Silver Mine in 1925, and on May 26, 1926, together with three other men, he founded Placer Development Limited with headquarters in Vancouver.

The company expanded with offices in San Francisco and Sydney, Australia, and operated placer deposits in BC, Alaska, the US, and Colombia before "hitting it big" on the Bulolo River in New Guinea. The World War II years were not good for Placer, but operations in Colombia helped it stay afloat.

During the war Banks worked for the Dominion Government and was responsible for shipping all military supplies from Canada to Britain. Probably as a reward for that work, he was appointed Lieutenant-Governor of BC and served from 1946 to 1950. In 1959 he retired from Placer and moved to Vancouver, but he kept homes in both Victoria and San Francisco.

The following is how Stan Hunter, a mining engineer who worked at the Emerald from 1949 to 1952, remembers Banks:

> Banks was the heart and soul of Placer, an admirable man. Although he was a multi-millionaire, he had a great personality. He was quiet, dignified, and treated everyone with courtesy, regardless of their position. Whenever a visitor came to the office, he would immediately instruct the secretary-receptionist Mrs. Currie, "Please set this man down and make him some tea." He always took time for visitors. It was a wonderful experience to see old man Banks performing. Yet despite his openness and friendliness, he was also very formal. No one addressed him by his first name, ever!
>
> Management came up twice yearly for a meeting with staff. Banks would have Ivor Phipps, the cook, put on a formal "army tea" in Lieutenant-Governor Fashion.
>
> The spirit of Banks was the spirit of the Emerald mine. The success of Canex and the success of Placer was due to Banks. He had the personality and the intelligence and knew how to use them. He was a very wealthy man who made his money out of placer mining, but he was approachable. He treated everyone the same. It did not matter whether you were the water boy or staff, he treated you respectfully. Yet when he wanted something done, you knew it, and you would not waste the company's money in doing it either.

Placer became a very large and successful company, and I attribute this to Banks. He did not try to manage the property, but managed the company instead. He was a firm believer in exploration and in the markets. He knew his markets, where they were and how large they were. He was an important mentor and has greatly influenced my life.

Banks was also receptive to World War II veterans, so the Emerald employed many of them. It paid off, for unlike me, they were stable, and looking to get their lives back to normal.

I think the mine's success was due to Banks assembling a first class team. They had people like Doug Little, Joe Adie, Al Lonergan—a good group. Doug Little later became vice president of Placer. Joe Adie was chief engineer for a while before moving on to other properties. They had an excellent group of young people there who stayed with them.

The Emerald was Placer's first lode mine, but others would soon follow. The Craigmont Mine at Merritt (copper) opened in 1960, followed by the Endako Mine at Fraser Lake (molybdenum) and the Gibraltar Mine near Williams Lake (copper). Craigmont had a mix of open pit and underground operations, but Endako and Gibraltar were strictly open pit mines. The company then went international and opened mines in the Philippines and New Guinea. With this rapid expansion, it became important to train people, and the Emerald Mine would become known as the Canex School of Mines.

Canex reopened the Emerald mine in early 1947 under the direction of G. M. Christie, the general manager. His original staff included H. M. Powell as mine superintendent, with T. B. Magee as his assistant, and Harold Lakes as a consulting geologist. The Emerald Tungsten was the first of several tungsten ore bodies that Canex

Harold and Barbara Lakes's first (tar paper shack) home at the Emerald. Late 1940s. Courtesy of Peggy Green (niece of Barbara Lakes).

Barbara Lakes—Christmas at the Emerald, late 1940s. Courtesy of Peggy Green (niece of Barbara Lakes).

mined on Iron Mountain. The others were Dodger, Feeney, East Dodger, and the Invincible. The Jersey was a large lead-zinc deposit that included several zones.

There are two general types of mining: development and production. Development work entails opening up new mining areas, which usually means driving "drifts" (horizontal tunnels) or "raises" (steeply inclined tunnels). A "prima donna" miner would always opt for development because it generally pays better. Development mining requires superior knowledge and organization, as well as harder work. Production mining, on the other hand, is drilling and blasting in already developed stopes.

In addition to the usual development, the new owners also had a lot of construction and repair work to do in order to bring the mine up to the desired standard. They set about repairing the tramline, overhauling the mill, and replacing equipment that had been sold. They upgraded the camp, began underground work, and drove raises from some stopes up to surface to improve ventilation. *(See box on next page.)*

By June the company was mining and milling the first tungsten ore and by the end of the year was processing 260 tons of ore per day. At first most of the ore was obtained from the mine dumps. Thus the mine was brought up to full production in an organized manner, and by the end of the year it was possible to take the bulk of the tungsten ore from underground.

The tungsten concentrator was located beside the Nelson–Nelway highway five miles south of Salmo, and the ore was fed to it via a 6,700-foot aerial tramway from the Emerald mine. Initially, extraction ran at about 60% WO_3, but this was soon improved to 70% under the direction of mill superintendent H. Grimwood, and by 1956, the concentrate would exceed 80% WO_3. The company also constructed a tailings pond between the highway and the Salmo River for storing the waste from the mill so that it would not enter the waterway.

In an article originally written in the 1990s for the *Northern Miner* magazine, Stan Hunter reminisces about the tramway and describes some of the difficulties the company had with it:

A tramline connected the mine on the mountain with the mill in the valley of the Salmon River. This installation had been constructed by Cominco during the Second World War under the stewardship of Pat Stewart and Ernie Rhyder.

One of the main features of the mine was the cable system that carried ore in buckets down the mountain to the mill at the rate of 400 tonnes per day. The wire system was subject to the

> ## NATURAL VENTILATION
>
> In BC the underground rock temperature in a typical mine is around 10°C (50°F) and varies only near the mine portal or at substantial depth. This is a comfortable temperature in which to work, because the miner is not likely to perspire too much, and yet it is not so cold that he would be uncomfortable, unless he sits around for too long. Rock temperature increases about 1°F for each one hundred feet of depth. The bottom levels at the Pioneer Gold Mine near Bralorne, for example, were 5,000 feet below the surface and hence about 100°F.
>
> One feature in many BC mines was that there were a number of different levels, with the ore fed from the various levels through ore passes down to the bottom level, from where it was hauled out to the surface. Both the top and the bottom level often had exits to the outside. This created a chimney effect, which facilitated great ventilation for most of the year. In the winter, when the air outside was colder than the air inside the mine, it entered the mine at the bottom level, then, as it was heated by the warmer rock, rose and exited at the top level. In the summer, when the air outside was warmer than the rock inside, it would enter at the top, fall as it cooled, and exit at the bottom. The greater the temperature differential between the air outside and the rock inside the mine, the faster the air would move through the mine. Only when the temperatures inside and out were about the same would there be poor or no natural ventilation.

vagaries of the weather. In winter, snow, hail and fierce winds assailed the installation, resulting in frequent shutdowns. In summer, violent lightning storms erupted across the mountain, sometimes striking tramline towers and rendering the system inoperable. Soon, it became company policy to shut down the trams during such storms, which hampered production. (Years later, the tram was shut down permanently in favour of a conveyor through the mountain that delivered a continuous supply of ore to the mill.)

The key component of this tramline was three miles of three-quarter-inch steel cable. The cable, to which were attached the ore buckets, was subject to great wear and tear, but it was never allowed to disintegrate. When sections of the cable appeared to be wearing, Sonny Burgess was summoned from his home in Ymir, south of Nelson.

Based on his recommendation, a replacement cable would be ordered. Sonny would then display his talents. The old cable was removed, and the new cable strung between the towers and across the mountainside. The critical part of the operation—and Sonny's specialty—involved splicing the two loose ends to form a continuous cable. First came the adjustment of the length and tension as the two ends were drawn together. Sonny then set to work fabricating the 58-ft. running splice, which involved separating and splicing together individual strands and the core of the cable. It required 10 to 12 hours of detailed labour by five or six men (and there was no second-best). Sonny performed masterfully, and always stayed to oversee the first run of the new cable.

The tungsten ore bodies were difficult to delineate because of their irregular shapes. One needed to use ultraviolet lamps to determine what constituted ore, and what did not, for the difference could not be seen by the naked eye. Furthermore, the deposits were in the shape of lenses that were thick in the middle but thinned out rapidly away from the body centres. Nor were these lenses continuous. The company therefore had to diamond-drill constantly in order to ensure that all ore was mined, while not diluting it with waste rock. It shouldn't come as a surprise then that Canex maintained a number of geologists on the payroll. To make the outline of the ore-bodies easier to grasp, the staff prepared glass models of the mine. Plotting the latest information from the diamond drilling and the mining on this model and keeping it up to date helped everyone to better understand it. Geological work was also done continuously with a view to finding both tungsten and lead-zinc ore extensions. Exploratory diamond drilling was initiated along extensions of the ore zones.

The price for tungsten in 1947 was higher than in 1943 when the plant was built to meet the war-time demand, and the company had a ready market for its product. As things turned out, Canex soon learned that producing an acceptable concentrate without too much of the scheelite going into the tailings was anything but simple. Much of the concentrates had to be shipped to a refiner in New York for treatment before the product was marketable. After five years of patient experimentation, the company was finally able to produce an acceptable concentrate that met specifications, without losing too much during the process.

Management was intent on modernization, and Canex was the first mining company in Canada to implement "trackless" mining methods. Instead of loading and hauling the ore in small rail cars pulled by locomotives, adits were driven that were big enough to permit the use of large diesel-powered equipment. There were several American mines already using these methods, but none in

Canada. When the uranium mines in Ontario shortly began going trackless too, they recruited some of their key people from the Emerald Mine.

During 1948, routine mining operations were carried out above the 3,900' and 4,000' levels, which were mined through to surface during the summer. Preparations were also made to extend the mining down to the 3,800-foot level. The mill continued to process an average of 250 tons of ore per day. One hundred and twenty-two men were employed, with one third of them at the mill. The high-grade concentrates were shipped to the Atlas Steel Company, but 600 tons of low-grade concentrates containing 13% tungsten $[WO_3]$ were stockpiled because at that time prices were plummeting and there was no market. (Specification called for concentrates to contain a minimum of 60% tungsten.)

The tungsten mine was closed at the end of the year, and the last ore was milled on January 12, 1949. However, just prior to that time surface diamond drilling on the Jersey lead-zinc property had provided encouraging results. This meant that Canex would be able to switch to lead-zinc mining and keep operating despite the collapsing tungsten market.

■ ■ ■

Switching over to mining lead-zinc in 1949 required a major transition. The company had to construct a new road to the Jersey claims 4,000

This pit in 1949 was initially expected to produce about 5,000 tons of lead-zinc ore, but it eventually produced ten times that amount. Courtesy BC Minister of Mines Annual Reports.

feet to the south of the Emerald—they had been extensively drilled the previous year. The concentrator had to be converted to process lead-zinc ore. The compressors were moved from the Emerald to the Jersey, and the power line was extended as well. The surface was stripped and an open pit started. That pit produced 85% of the ore mined that year, and by March the mill was handling over 300 tons of ore per day.

There would be several more major transitions in the future: construction of a new mill and related facilities when the company resumed the mining of tungsten in 1951, the closure of the tungsten mine in 1958, and the switchback to tungsten after ending lead-zinc production in 1970. The ability to make these changes is the reason the mine could operate continuously from 1947 to 1973.

Lead-zinc townsite Quonset homes. Denis Hartland says there were 17 families living here. Courtesy of Ron Erickson.

In 1950 Harold Lakes took over as the mine manager. Even before that time his work as a geologist had been crucial to the company's success. Lakes was born in Colorado in 1886. Before coming to Canada, he had worked in Idaho, Mexico, New Mexico, and Colorado. In Canada, he was voted "Mining Man of the Year" at least once. No one seems to have a record of where Harold and his brother Arthur obtained their degrees, but Beth Simmons of the Community College of Denver, Colorado says: " . . . I think there's every chance that they earned their certificates through the ICS *[International Correspondence Schools]*."

Lakes first came to the Nelson area in 1914 to relieve his brother Arthur Jr. at the Ymir Wilcox mine and during the next five years was in charge of several properties in the Ainsworth–Sandon area. He came to Salmo in 1919 and was in charge of the Mother Lode and the Nugget at Sheep Creek, as well as the HB and the Reeves McDonald mines near Salmo. He was generally well liked, and Marie Adie, wife of geologist Lawrence (Joe) Adie, describes him as a "diamond in the rough kind of guy."

Under Lakes's management a new road was constructed from the Jersey mine down to the mill (the "south road"), so that the ore could be trucked directly to the mill and bypass the aerial tram, which was sometimes plagued with troubles, especially during high winds. The milling was raised to 500 tons per day as well. The Jersey ore zone had by now been proven over an area 5,000 feet long by 1,500 feet wide. The low-grade tungsten concentrates previously stockpiled were shipped to London, England.

■ ■ ■

In 1951 the Korean War was at its height, and tungsten, no longer available from either China or Korea, was again in great demand. By May 1st Canex had paid off the $900,000 owed to the government for the mine. Now the government repurchased a block of claims from Canex that included the Emerald tungsten ore body and a part of the developed Dodger for $328,000. Canex agreed to build a 250-ton/day mill at government expense and to mine tungsten ore on a fee basis. It built the mill and an independent camp, including residences, near the Emerald 3,800 adit and began milling tungsten ore in December.

Initial development in the Dodger started on the north end of the zone. A 14′ x 14′ adit was driven downgrade from the 4,400-foot level through part of the government-owned block. A second adit (crosscut) at the

---

IMPORTANCE OF MINING

Few of us realize the extent to which we depend on mining products in our everyday lives. The following are summaries of the principal uses of zinc, lead, tungsten and molybdenum, the four metals mined at Canex.

Zinc
Zinc is a metal that melts at 419°C (787°F), and has been in use for at least 2,000 years—the Romans used it to make brass. It is generally known that zinc is used to galvanize steel so that it won't rust, and that galvanized products are used in drainage culverts. Today we may not use much of it in brass, but zinc has a myriad of other applications. Our lives would be quite different without it.

In addition to its use as an anti-corrosion coating, zinc is used in die-casting, construction materials, pharmaceuticals, and cosmetics. It is also a necessary nutrient for humans. A number of studies indicate that many of us are deficient in zinc. Zinc, like some other metals, can be recycled indefinitely.

Lead
For many this metal has a dirty name. It was formerly associated with leaded gasoline, water pipes, and paints, all of which caused health problems. Working in a lead smelter often took its toll on the workers, who sometimes became "leaded." It was, until recently, used in ammunition (bullets and pellets), but has been discontinued for fear of poisoning waterfowl, which ingest shotgun pellets to help them digest their food.

Lead is now used mostly in lead-acid batteries, without which our cars would be quite different. It is also used as an insulator for radioactive materials and for piping in nuclear reactors. It, too, can be recycled—we all pay a fee for this when we purchase a car battery. Its melting point is 327.46°C (621°F).

4,200-foot elevation was started to intersect the zone 4,000 feet to the south of the Dodger 44 portal. These workings were ultimately connected, and mining was carried out using diesel trucks and loaders.

During 1947 a new tungsten ore zone south of the main camp had been discovered, diamond-drilled, and partly stripped. It was referred to as the "Feeney" and lay west of the Dodger zone and 700 feet north of the Emerald portal. An adit was started at the 4,035′ elevation and was driven several hundred feet by the end of 1951. While this was going on, the company had through diamond drilling continued to expand the proven Jersey lead–zinc ore bodies through a zonal length of more than 6,000 feet. In September 1951 the company placed the reserves of lead-zinc ore at 7½ million tons. John Waldbeser might have turned over in his grave, had he known that he had missed the big Jersey ore body by only 40 feet.

The Feeney ore zone was named after the man largely responsible for its discovery, and here is what Stan Hunter wrote about him:[5]

IMPORTANCE OF MINING—TUNGSTEN

Tungsten has been known about for over 200 years, but it was not until the middle of the twentieth century that it became heavily used in industry. Canada is second only to China in known tungsten reserves.

It is by far the strongest of all metals. It is also the hardest. It has a number of unique properties that make it suitable for use where its hardness, tensile strength, and electric conductance properties make it a popular choice for a wide range of applications. It is extremely heavy and has a density of 19.25 g/cm³ (almost twice the weight of lead) and melts at 3,422°C, (6,192°F.), the highest of any metal, and the second highest of all the elements after carbon.

Over 60% of tungsten production is used in "hard metals" applications, e.g. tungsten-carbide rock drill bits, high-speed drill bits, metal cutting tools, machine tools, etc. It has a hardness approaching that of diamonds.

Twenty one percent of production is used in steel alloys where it has the highest tensile strength of any metal in applications where temperatures exceed 1,684°C (3,000°F). Its anti-corrosive characteristics make it ideal for alloys exposed to harsh environments, e.g. sour gas pipelines. Tungsten alloys are also used extensively in airframe structures for airplanes and for turbine blades.

Eleven percent of it is used in filaments for lighting products, electrodes, welding applications, etc.

Eight percent of production is used in a host of other applications. A few examples include heat sinks in computer cooling units, magnetrons in microwave ovens, lasers, paint pigments, catalytic products, jewelry (non-scratch rings), etc.

---

5 Excerpted from a Stan Hunter article published in the *Northern Miner* in the late 1990s.

Joe Feeney and I first met at the Emerald mine, near Salmo, where H. L. Batten retained him to explore for mineralization on the Iron Mountain claims. *[Batten worked for Placer Development Ltd. when it first began searching for lode mines.]* Feeney prospected that mountain in meticulous detail. He made hundreds of surveys and broke tons of rock. Each day, he conferred with Harold Lakes, the geologist, on his observations, and Batten listened to Feeney with the deepest interest and encouragement. Feeney finally convinced Batten and Placer Development to conduct a major surface drilling program over the Jersey claims and on the adjacent tungsten zone. The results were spectacular in that Placer ventured into lead-zinc production on the Jersey and greatly expanded the tungsten operation.

With the large growth of ore reserves, the company divided the mining into three distinct operations, each with a superintendent, engineer, and geologist. These were the Jersey lead-zinc mine, the Emerald tungsten mine operating the government-owned ground, and the Dodger tungsten mine (including the Feeney ore zone). With the main camp as a point of reference at the 4,070′ elevation, the Jersey mine was 6,000 feet to the south, the Emerald mine and mill were about 3,600 feet to the southwest at 3,860′ elevation, and the Dodger mine was 2,500 feet northeast at 4,400′ elevation. The Feeney tungsten mine was south of and adjacent to the main camp. The lead-zinc concentrator, formerly the tungsten mill, was on the Nelson–Nelway Highway five miles south of Salmo, and the ore was transported to it by tramline and by road from the mine. *[See Figure II.1 for a general layout of the property.]*

During 1951, the Jersey mine operations were expanded to 800 tons per day. The ore was crushed near the portal and was trucked directly to the mill or to the head of the tramline near the Emerald portal. The company overhauled the tramline and increased its capacity. By the end of the year, there were more than 600 men employed in all operations.

■ ■ ■

From the foregoing it is clear that Canex was a modern company, being proficient at running the mine, producing ore, and adapting to changing market conditions. It had a "modern" outlook in other respects as well and knew how to take care of its employees. For instance, the company

> **IMPORTANCE OF MINING—MOLYBDENUM**
>
> Molybdenum (or "moly" as it is often called) is a soft, graphite-like mineral, which registers 1 on the Mohs scale of hardness (similar to talc), but in its pure metal form it is very hard. It melts at a temperature of 2,623°C (4,730°F), which is 1,111°C higher than the melting point of steel and the fifth highest of all the metallic elements. When alloyed with steel it creates a product that is extremely hard, strong, and resistant to corrosion. The iron and steel industry accounts for 75% of all usage. In the United States, 60% of its production is used in stainless steel and steel alloy production. Stainless steel is much used in water distribution systems and food handling equipment as well as in chemical processing, hospital, and laboratory equipment. Molybdenum alloys are used in bearings, machine tools, gas transmission pipelines, auto parts, and as super-alloys for gas turbines, metal working dies, etc., to name just a few applications. The metal has important lubricant qualities as well. It is used in synthetic oils and greases as well as being an additive to premium motor oil. It is also used as a catalyst in paint pigment, smoke and flame retardants, space vehicles, and many other purposes, far too numerous to list here. It is also an essential nutrient that is important for the health of plants, animals, and humans.

implemented and contributed to mandatory health insurance—a rarity at the time. During 1951 the company began building and/or renovating dwellings for its workforce on a larger scale.

This step cannot only be attributed to trying to reduce the commute for the employees to the remote location—it wasn't all that far to Salmo and Ymir—8 and 15 miles respectively (13 and 24 kms)—and in fact, single men continued to live in the surrounding communities, with a few commuting even from Nelson and Fruitvale. The bunkhouses at the mine, where single workers had lived for some time, were phased out once construction tapered off.

Now, around 1951, a real townsite sprang up below and west of the Jersey mine buildings and extending towards the Emerald camp. Eventually there would be close to 150 family residences, many of them duplexes, with most of them being at the Jersey, a few at the Emerald, and a few at the lead-zinc mill down in the valley.

This move to create residences for families perhaps points to Canex's intention to create something like loyalty of the employees. The Emerald never became quite the company town that the interior of British Columbia saw in other areas. One reason was perhaps that by the 1950s families began to have their own vehicles, which made driving to close-by communities easy. For instance, for the bulk of their grocery shopping most of the employees went to Salmo and Ymir or even to Nelson and Trail, usually on days off after the semimonthly

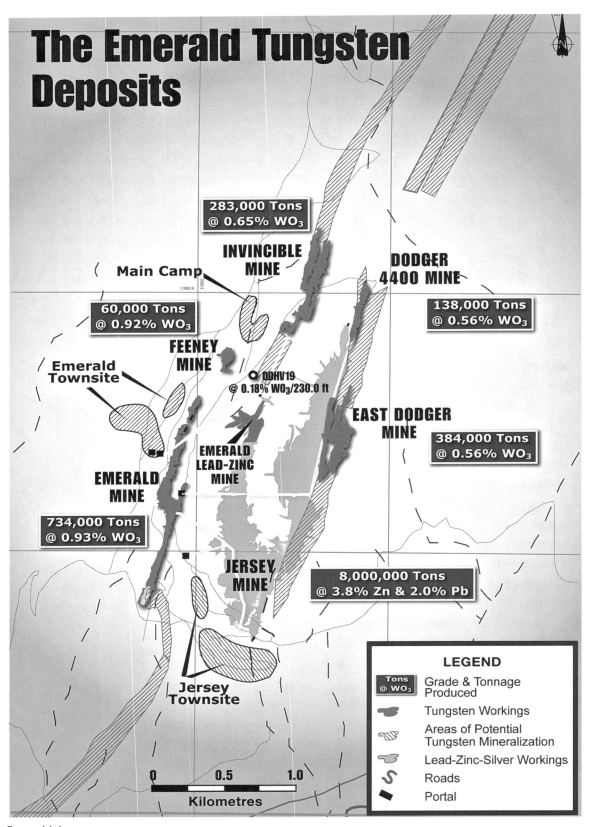

Figure 11.1

40 / JEWEL OF THE KOOTENAYS

> **ROTTEN EGGS**
>
> Much is often made of the quality and quantity of food in mining, logging, and construction camps. In an article published in the *Northern Miner* magazine many years ago, Stan Hunter has fond memories of the food at the Emerald Mine. My recollection of it is quite different.
>
> In all fairness, I must admit that it was a very different camp in 1952 when I was there, from what it had been two or three years earlier. During Hunter's time there was a much smaller crew to feed. During 1951 the workforce grew quickly to 600 employees, and during 1952 the crew even grew to a force of over 900—probably three times the size it had been during Hunter's time.
>
> I don't remember too many details of the food in the cookhouse except for the eggs—they were awful. I believe that the mine was buying old "storage" eggs. Whatever the case, they were so old that the yolks in boiled eggs would harden, while the whites stayed runny (and slimy). What's more, they emitted a nauseating odour.
>
> In other mines where I had worked, it was the custom for the men to make up their own lunches from a table laden with bread, butter, deli meats, cheese, jam, etc. One could be creative about one's lunch. During my stay at the Emerald, the kitchen staff prepared the lunches.
>
> I found the breakfasts nauseating and the lunches drab. It was not long before I was able to move into Mrs. Gibbon's boarding house in Salmo. It cost me more money, and I had to commute as well, but for me it was still a big improvement over the camp.

payday. The mine did have a commissary—a store where gloves, boots, cigarettes, and other items could be bought, as well as in an emergency a few groceries, but the prices were much higher than in town. There was an elementary school at the mine for students up to and including grade six. From then on, the children attended the high school in Salmo. The bus that brought the miners up from Salmo took the students down to the school. The nearest hospitals were in Nelson and Trail—35 miles (56 km) away, but Salmo had both a doctor and a clinic. There was also a church in Salmo; still, as the interviews show, the common feeling among the residents was one of belonging and of equality. "It didn't matter whether you were staff or hourly paid," says Patricia Rose, who grew up at the mine, "you got your house assigned on the basis of family size."

With the work week down to 40 hours in the mid-fifties (as opposed to the 70 hours when the Emerald first opened) and no time spent on commuting to work, people at the mine had a lot of leisure time. It is here where Canex contributed substantially to the wellbeing of its employees and their families. For example, the company provided a recreation room, where weekly films were shown. And the policy adopted by Canex to contribute equipment and materials for recreational facilities

This photo of the lead-zinc tailings pond designed by Tony Triggs is from the 1956 Minister of Mines Annual Report.

This photo of the lead-zinc tailings pond was taken from the south road by John Bishop in 1964. It shows a well developed tailings pond. Courtesy of John Bishop.

The lead-zinc tailings pond was closed in 1970. In 1971 it was growing new grass. Courtesy of Sultan Minerals Inc.

while calling on residents to provide volunteer labour did more than reduce costs: it fostered a true community spirit. The projects built under this policy included a swimming pool (heated by the water used to cool the big mine compressors), a curling rink, and an ice rink. Other popular sports included badminton in winter, as well as horseshoes, softball, fishing, and hunting in summer and fall. Fishing in the Nelson district streams and rivers opened July 1st, but the lakes were open all year. After the mines stopped their practice of dumping tailings into the creeks and rivers, fish populations recovered and fishing became an important pastime. Many miners also hunted.

Another important recreational event at the Emerald was the picnic the company hosted every summer for its workforce at Sullivan Lake in Washington State—a site that was used by employees and their families at other times as well. There were also other local community picnics, which featured mining-related contests such as drilling and "hand-mucking" for men and foot races and the like for women and children. The individual mines usually hosted picnics of their own as well with similar events. During one such picnic I teamed up with another miner, Wally Bunka, and won the log sawing contest for crosscut saws. We were quite pleased with ourselves that we beat the loggers at their own game.

Recreation for single men, of which there were many, included other activities as well. The local hotels (two in Salmo and one in Ymir) played an important part in this, as their beer parlours or pubs were sometimes the only legal source of beer (government liquor stores being few and far between). Dave Martin, son of Mike Martin, then the owner of the Ymir Hotel, told me that his father sold more beer (illegally) on Sunday than during the entire remainder of the week. The price for a case was $3.65, but his dad had instructed

The Ymir Hotel as it looks in 2007. Mike Martin's son Dave upset his father when he painted it purple during his father's absence. It was referred to for many years as the Purple Peanut.

Dave to tell the customer that he had no change for the $4.00 usually proffered. Mike Martin "helped" his customers in other ways as well: he often advanced money to certain patrons to tide them over to payday. It would not be long after payday before the individuals would be back for another loan.

At the company townsites, the Emerald as well as at the others, poker and bridge games went on but usually for only small stakes. It was a different story in Nelson, where one licensed poker den stayed open around the clock, at least on weekends.

There were two brothels in Nelson during the fifties—an important institution for many of the miners of the time. Some single miners would even drive down to the Mustang Ranch at Priest River, Idaho, to satisfy their needs.

Many of the men commuted to the mine by bus from Salmo, but during heavy snowfalls that was sometimes a problem. The north road to the mine was shorter but steeper and tougher to get up than the south road, which had 22 switchbacks. The crew was always able to make it down though. The mine had provided the bus service without charge until after the labour strike in 1959, when Gerry Gordon began charging the workers $0.10 each for their rides.

Of course, the flourishing community at the Emerald depended entirely on the company's fortunes, and once the mine was shut down completely, no one continued to live there.

JEWEL OF THE KOOTENAYS / 43

This 250-ton mill was built at government expense in 1951 to process tungsten from the government repurchased property (on a fee basis). The company also constructed an independent camp as shown, and a six-mile-long 10" water pipeline from Lost Creek. However, when Canex discovered important additions to the tungsten reserves, it repurchased the government holdings in October 1952 and by year-end had increased the mill capacity to 650 tons per day. Courtesy Colin Brown.

■ ■ ■

1952 was a momentous year for Canex. Development reached fever pitch as the company, in addition to its other construction work, embarked on a new transportation system that included a new dual-purpose crusher and a novel conveyor system. It also increased the mill capacities, drove new adits for the Dodger ore body, and repurchased the government assets it had sold the previous year.

It must have been quite a load of responsibility for the 66-year-old Harold Lakes, for he suffered a stroke that year and retired to Nelson, where he died on Christmas day four years later. Gerry Gordon took over as the general manager, a position he would hold for several years before moving into Head Office in Vancouver. He was assisted by a large staff of superintendents. *[Refer to Table II.1].*

The company repurchased the tungsten holdings from the government on October 1st, 1952, and increased the tungsten mill capacity to 650 tons per day. By year-end the Jersey lead-zinc mill was processing 950 tons per day. Since the start of mining operations in 1949, the Jersey flat-lying ore bodies known as A, B, C, and D had been mined in an area 1,800 feet long and 800 feet wide.

The aerial tramway from the mine to the lead-zinc mill had given a great deal of trouble, so work began on a number of drifts and raises for the installation of a 7,000-foot-long conveyor system that was completed the following year. An underground crushing system comprised of a jaw crusher and a secondary crusher was installed. This crusher was designed to draw either tungsten or lead-zinc ore as required. The conveyor system (the longest of its kind in the world then) was designed to send the ore from the crusher to the appropriate mill. It included several large underground ore storage bins that ensured that neither mill would run out of ore whenever the system was providing feed for the other mill. The company was thus able to provide both concentrators with a steady supply of ore from a single crushing system. This system replaced the unreliable aerial tram and reduced both crushing and transportation costs substantially.

An open pit was started on the Emerald D zone. At the Dodger Forty-Four, a 14' x 15' adit was driven south for 1,050 feet. From the Dodger 42, a similar-sized adit was driven east for 2,500 feet to eventually intersect the Dodger 44 adit. All work in the Dodger was by "trackless" mining methods using large diesel-powered equipment.

By January 1953, with all of the expansion work in progress, the workforce had peaked at 900, but from then on, as construction was completed, it declined rapidly, until in December it was down to a more manageable crew of 350. The concentrates

Tungsten camp. Mid-1960s. Compare this photo with the Colin Brown picture from 1951: the old bunkhouse is gone, but the mine dry is still there. A shop now sits in front of the mill. No. 1 and 2 conveyors together with the transfer house can be seen in front of the shop, which is also recent. Courtesy of John Bishop.

Skipway at bottom of No. 6 conveyor. Courtesy of Sultan Minerals Inc.

Conveyor junction and tungsten mill. Courtesy of Sultan Minerals Inc.

Table II.1: Mine Management

MINE MANAGERS

Iron Mountain Limited

1905 – 1925	John Waldbeser
1938 – 1942	Harold Lakes

Wartime Metals Corp.

1942 – 1943	E. E. Mason

Canadian Exploration Ltd.

1947 – 1948	G. M. Christie
1948 – 1949	R. E. Legg
1950 – 1952	Harold Lakes
1952 – 1961	G. A. (Gerry) Gordon
1962 – 1964	R. (Bob) G. Weber
1964 – 1968	C. E. Brown
1968 – 1974	Ed Lawrence

ASSISTANT MANAGERS

	R. S. Douglas
1952	G. A. Gordon
1953 – 1954	G. W. Walkey
1955 – 1960	J. D. (Doug) Little

CHIEF GEOLOGIST

1950	Clive Ball

MILL SUPERINTENDENTS

1947 – 1950	G. H. Grimwood, G
1952 – 1955	G. Walkey, G

MINE SUPERINTENDENTS

1955	J. D. Little, G
1956 – 1960	H. A. Steane, G
1962 – 1963	R. W. Gould, G
1952	V. McDermot, T
1953 – 1959	R. McLeod, T
1951	R. Mason, L–Z
1952 – 1960	E. A. Erickson, L–Z
1964 – 1965	B. Wilson, L–Z
1966 – 1967	D. A. Knight, L–Z
1968 – 1970	A. Filyk, L–Z
1970 – 1973	???, T

PLANT/GENERAL SUPERINTENDENTS

1957 – 1960	C. M. McGowan

MINE SUPERINTENDENTS

1947	H.M. Powell, T
1948 – 1950	J. B. Magee, T
1951	G. A. Gordon, L–Z
1952	W. F. Atkins (Feeney), T
1952 – 1953	J. D. (Doug) Little, L–Z
1952 – 1955	H. (Hub) Maxwell, T
1954	A.D. McCutcheon, L–Z
1955 – 1957	D. N. Hogarth, L–Z
1958 – 1961	R. (Bob) G. Weber, L–Z

LEGEND: G = General, T = Tungsten, L-Z = Lead-Zinc

were sold to the US Government through long-term contracts negotiated by Simpson, the managing director of Placer.

Until the new crushing-conveying system was in operation in July 1953, the mine trucked the tungsten ore directly to the mill from the Dodger. After that, the new crushing plant processed both the lead-zinc and the tungsten ores and sent them to the concentrators by conveyor. The tungsten concentrator milled an average of 9,000 to 10,000 tons of ore per month from all sources, with 2,000 tons of that coming from the Dodger by autumn. The difficulties in producing a marketable product were gradually overcome by fine-tuning the mill. A roaster was also installed to produce tungstic oxide from some of the concentrate. This product fetched a better price than the scheelite. The mine was still a beehive of activity, but quiet as compared to a year earlier.

One noteworthy event was the kitchen-staff house at the Jersey burning down[6]. It contained 18 rooms as well as the kitchen, lounge, and dining room. Despite the efforts of the fire crew, the fire leveled the building in just 15 minutes. It had been built just a year earlier at a cost of $75,000 and in addition to this equipment and personal possessions valued at $25,000 were lost.

In 1954 a labour strike was averted when the company reduced the work week to 40 hours as well as giving the men a small pay raise.[7] The Jersey ore above the 4,200' level was removed by trackless mining, and that below, by conventional methods. A. D. McCutcheon took over as superintendent of the lead-zinc operations, and that mine now extended 5,000 feet to the north and was up to 2,000 feet wide, with the ore body being up to 60 feet in depth but averaging 12 feet. It sloped from north to south at 10° to 20°.

At the Emerald Tungsten, ore from near the surface was recovered in large open pits. A winze (shaft) was sunk 755 feet at a dip of −32° below the horizontal. The Feeney had been stoped from the 3,800' level through to surface and produced more ore than expected. Monthly milling rates for tungsten were now from 10,000 to 12,000 tons per month, and for lead-zinc, about 30,000 tons per month.

During 1955 the Emerald ore body continued to produce most of the tungsten ore. The winze that had been sunk from the 3,800' level for a slope distance of 1,390' began to produce ore late in the year. These ore bodies tended to be smaller but richer than those higher up. The Feeney had produced over 60,000 tons, but was mined out. The drift connecting the Dodger 42 to the Dodger 44 was enlarged to 14' by 14' to allow diesel trucks to haul from the Dodger 44 to the ore-pass outside the Dodger 42. This ore-pass fed directly to the crusher at the 3,800-foot level. The Jersey mill, operating at half capacity during that year, averaged 30,000 tons of lead-zinc ore per month.

During 1956, the Emerald ore body for the first time produced less than 50% of the tungsten ore mined that year. A considerable amount of this ore was from open pits on the surface, where outcrops had been stripped for 1,000 feet. The winze had been sunk to the 2,730-foot level.

Harold Lakes feeds his ducks near his home on Kootenay Lake, 1950s. Courtesy of Peggy Green (niece of Barbara Lakes).

Peggy Green, 2007. Peggy, born on 6 April 1923, is a niece of Barbara Lakes. I first visited her early in 2007 in her apartment in Nelson, but she had most of the things I was interested in at her lakeside home just north of Balfour. When I visited her there in August, she provided me with most of the Lakes's photos I have used in this book.

---

6   *Nelson Daily News*, Nov. 18, 1953.
7   *Nelson Daily News*, May 15, 1954.

No. 1 conveyor transfer house. Courtesy of Sultan Minerals Inc.

2A conveyor to tungsten mill. Courtesy of Sultan Minerals Inc.

Transfer from No. 2 to No. 2A and to No. 3 conveyors. Courtesy of Sultan Minerals Inc.

This shaft closely followed the ore body, which plunged at approximately the same slope as the shaft. An extension of the Dodger 44 ore body was found in the east wall of the adit. Irregular-shaped ore bodies were mined above the main drift and hauled by diesel truck to the ore-pass at the Dodger 42 portal. The tungsten mill continued to expand its output slowly and handled an average of 17,300 tons per month for the year. Diamond drilling had detected the Invincible ore body, north of the Feeney and west of the Dodger 44.

At the Jersey there were now six ore zones known as A, B, C, D, E, and F. They were close together at the south end, but more separated towards the north. They varied greatly in thickness, with the A zone over 60 feet deep in places, whereas the E and F zones were only from 8 to 10 feet thick. The A zone produced most of the ore, and the mill, working at half capacity, averaged 31,500 tons per month.

In 1957, the Emerald ore body again produced most of the tungsten ore, the majority of which came from surface pits extending for 1,500 feet, where the outcrops had first been stripped. The average crew size for the past two years had been about 350. The ore on the east side of the Dodger 44 adit was mined to surface, and most of the known ore in the Dodger had been removed, as had most of the ore above the Dodger 42. The indicated tonnage of the Invincible ore body was 386,000 tons grading 0.83% tungstic oxide. A start was made on a 900-foot vertical shaft to mine this ore, but was stopped in October because of low prices. The tungsten mill averaged 15,000 tons per month. At year-end, only the Emerald was still producing tungsten ore, the others having been mined out.

The A zone in the Jersey was now fully developed and extended over a distance of 5,000 feet. The B, C, and D zones had each been developed for 2,000 feet. The E, F, and G zones lay further to the east and were not distinguishable at the south end. Seventy percent of the Jersey ore was now mined by trackless methods, and the rest of it by conventional means. The Jersey mill operated on a four-day week and averaged 2,000 tons per day during 1957, with the concentrates being shipped to Kellogg, Idaho.

The tungsten operations were shut down on July 31, 1958; all developed ore had been mined out above the 3,800' level in the Emerald. *[The long term contract with the U.S. government had probably expired as well]*. The Jersey mill handled 32,000 tons per month, with grades of 4.2% zinc and 2.3% lead. Concentrates continued to be shipped to Idaho. Ore in the E, F, and G zones, being thinner, was mostly mined by conventional means as opposed to trackless.

■ ■ ■

During the 1959 to 1969 decade, Canex operated only the Jersey lead-zinc mine. It was during this period that the mine became known as the "Canex School of Mines," for the company provided most of the key staff for the Craigmont mine when it opened in 1960/61. It also provided some staff for the Endako and Gibraltar mines.

Drilling pillars from a raise machine, mid-1960s. Courtesy of BC Minister of Mines annual Reports.

The safety program paid off. The company suffered only one lost-time accident per year for three consecutive years. In 1960 it won the Ryan Trophy for the best safety record for a medium-sized mine. Courtesy of Pat and George Sutherland.

In 1959 the Jersey A zone was further developed by sinking a 6' by 9' winze at −32° to 240 feet below the 4,000' level. The main trackless haulage was via the Dodger 4,200' level and from the track mine on the 4,000' level. Ventilation was improved by connecting a raise to the Dodger 4,400' level. Lead–zinc concentrates were shipped mostly to the smelter in Idaho.

In 1960 the main haulage was via the Jersey 42 and Dodger 42 adits. Seventy percent of the loading was done with front-end and overhead loaders, and the ore was hauled with eight-ton Dumptors to the ore-pass that fed the crusher on the 3,800' level. Open stope mining was done with jacklegs, slushers, and a three-machine jumbo mounted on caterpillar tracks. A truck-mounted Giraffe was used full time for checking and scaling the backs (roofs) in the stopes. In the track section the ore was hauled by diesel-electric locomotives. Ore from below the 4,000' level came from the winze.

Because some of the ore zones were so wide, temporary pillars of un-mined ore had to be left in place to support the back (roof). In the A zone especially, where the ore body was very thick, this resulted in a lot of ore being left. The intention was therefore to remove these pillars once the zone had been mined to its extremities. Then beginning with the pillars farthest from the portal, they would be drilled, blasted, and their ore retrieved, as the operations retreated toward the mine entry point.

The pillar recovery program in the D zone was launched in the fall of 1960. An Alimak raise-climbing machine was used in the A zone to drill off pillars, which now ranged up to 90 feet in height. Special explosives were used in the top two rows of holes to leave a smooth back. The plan was to scrape the ore to a loading point so that no one would have to work underneath the unsupported back.

Almost two miles of raises and drifts were driven during the year. Ventilation was maintained by three large electric fan systems. Zinc concentrates for the last two months were shipped to the Trail smelter for the first time. The tailings pond was raised, and tests were made to determine the best cover to seed and/or plant on it.

There was a "game of musical chairs" in 1961 as some staff moved to the Craigmont Mine or into Head Office. Jack W. Robinson replaced Bob Weber as mine superintendent, when Weber became the property superintendent and R. W. Gould took over as mill superintendent.

A major event was the introduction and use of AN/FO explosives by year-end (ammonium nitrate prills factory mixed and sold as "Amex"). This innovation produced substantial cost savings because the Amex sold for a small fraction of the price charged for dynamite. The Amex had to be blown into the blast-holes using air pressure and was not suitable for every situation. Furthermore, since no one had used it before, it was naturally prudent to proceed with some caution at the start. If the Canex miners were like those in other mines, they would have been resistant to adopting its use for it was a bit of a pain to handle. Dynamite could in any case not be totally eliminated, for it was needed to detonate the Amex.

The trackless mine provided 65% of the ore. Three pillars in the track mine were blasted during the year. One pillar in the A zone was drilled with the use of an "Alimak" raise-climbing machine, but not yet blasted. There was just one lost-time accident of over six days, and the company won the Ryan trophy for the best safety record for a mine it its class.

During 1961 zinc concentrates were shipped to Trail for the first six months, after which they were again shipped to Idaho, as were all lead concentrates for the year. This year also brought a change to the Metalliferous Mines Act, the ramifications of which would be felt in the Emerald as well. In keeping with increased concerns for safety as well as heightened awareness of the legal aspects of the mining industry, all shift bosses would in the near future have to be certified and pass exams to test their knowledge of the act.

In 1962, Weber took over as the general manager, Robinson was still the mine superintendent, and Gould remained as mill superintendent. During that year, the trackless mine produced 83% of the ore. AN/FO accounted for 55% of all explosives used. CIM consultants performed stress testing on the pillars, using strain gauge disks. They found that pillar removal did little to change the stress patterns. It appears that the rock was generally so strong that the pillars were not needed for supporting the back. Just the same, no prudent management could have chanced men working in such wide stopes without pillars to support the back to ensure the safety of the crews. Just a year earlier, on May 5, a huge rock slab had fallen to the stope floor. Fortunately no one had been hurt in that incident, but it must have served as a reminder to management that the risk of fatalities was unacceptable.

Mill production remained stable with all concentrates being shipped to Idaho. There was only one lost-time accident of more than five days for the year.

In 1963 the pillar recovery continued and had been standardized using wooden stagings. No reason is given in the government reports for not using the Alimak raise machine—perhaps all the high pillars had been drilled by then. Mining in the tracked section was completed, and rail, pipe, and other services were removed. The use of AN/FO increased and now accounted for 70% of all explosives used. A surface exploration program started in 1962 was completed.

In 1964 Bob Weber was killed in a CP Air plane crash near 100 Mile House. Apparently someone had placed a bomb on the plane. Everyone including me was shaken by the news—I knew Bob Weber as a friendly, personable fellow from 1949 when he was an engineering student working at the Paradise Mine while I was there. C. E. Brown replaced him as the mine manager.

Two-thirds of the zinc concentrates were shipped to Trail in 1964.

In 1965 the lead concentrates were shipped to Kellogg, and 47% of zinc concentrates were shipped to the Trail smelter, the rest to Anaconda in Black Eagle, Montana.

D. A. Knight took over as mill superintendent in 1966. The company shut down the mill for a three-week holiday in July but still processed a record amount of ore and was approaching 40,000 tons per month by year-end. Most mining was still done with jacklegs, but the jumbo was used on large headings. The company acquired two new Trump Giraffes: a 40-footer and an 85-footer, as well as a Wagner scooptram and a 966B loader. Ore reserves at year-end were slightly over 606,000 tons. Ten people successfully wrote shift-boss exams.

Nineteen sixty-seven saw record production of lead-zinc ore, and reserves at year-end were down to slightly more than a half million tons. The company operated 31 pieces of diesel equipment underground. A smelter strike in the US resulted in zinc concentrates being stockpiled for half of the year.

In 1968 Ed Lawrence took over as the mine manager and A. Filyk as mill superintendent. Reserves at year-end climbed to over 630,000 tons. Three new areas were developed: the J zone, the east F zone, and the 33D zone

Refurbished tungsten mill, circa 1970. Courtesy of John Bishop.

below the old track mine. Pillar extraction in the north A zone was started. Production for the year exceeded a half million tons. Two new RD-13 Euclid trucks and a D6B dozer were purchased during the year. In a new twist, 11,500 tons of zinc concentrates were smelted and returned to Canex, which disposed of them on world markets. Several falls of ground took place in heavily fractured areas of the Jersey.

Diamond drilling on the Invincible ore body during the year had proved up over 300,000 tons grading 0.70% tungstic oxide.

During 1969, the mill treated 2,500 tons of ore per day on a five-day week. As ore reserves dropped, an increasing amount of ore came from pillar recovery. The Jersey closed early the following year. Table II.2 summarizes the work done in the Jersey lead-zinc mine.

■ ■ ■

Once again, after many years of producing only lead-zinc, demand and favourable world market prices induced the Canex management to switch back to tungsten. The 6,000-foot Invincible adit was begun in 1970, at a −16° grade, and completed the following year. The Emerald tungsten mill was rehabilitated, and production started in October 1971. By year-end the mill was processing 430 tons per day on a trial

1966 – Voters List Nelson-Creston District 207
No. 37 CANADIAN EXPLORATION MINE POLLING DIVISION

1. Ablett, Moraine Margaret, housewife, Canadian Exploration Ltd., Salmo.
2. Adolphson, Mae Lucille, housewife, Salmo.
3. Adolphson, Olof Gunnar, mine captain, Emerald Mine, Salmo.
4. Anderson, Clarence Perry, foreman, Emerald Mine, Salmo.
5. Anderson, Hazel, housewife, Emerald Mine, Salmo.
6. Ash, Alice Grace, housewife, Canadian Exploration, Salmo.
7. Ash, Richard, paymaster, Canadian Exploration, Salmo.
8. Backler, Barbara Jeanne, housewife, Canadian Exploration Ltd., Salmo.
9. Backler, Brian Edwin, engineer, Canadian Exploration Ltd., Salmo.
10. Boys, Bernice Alice, housewife, Salmo.
11. Boys, William Roy, mechanic, Salmo.
12. Brown, Colin Edgar, mine engineer, Salmo.
13. Brown, Eileen June, housewife, Canadian Exploration Ltd., Salmo.
14. Clayton, Frank, mechanic, Canadian Exploration, Salmo.
15. Clayton, Helen May, housewife, Canadian Exploration Ltd., Salmo.
16. Clayton, William Gordon, machinist, Salmo.
17. Coley, Wenda Beth, housewife, c/o Canadian Exploration, Salmo.
18. Colwell, Coburn Irvine, purchasing agent, Salmo.
19. Colwell, Rhea Pearl, housewife, Emerald Mine, Salmo.
20. Darychuk, Bella, housewife, Emerald Mine, Salmo.
21. Darychuk, Nick, miner, Salmo.
22. Davis, Angus Ward, secretary, Canadian Exploration Ltd., Salmo.
23. Davis, Helen Arline, housewife, Canadian Exploration Ltd., Salmo.
24. Dion, Barbara Ruth, housewife, Canadian Exploration Ltd., Salmo.
25. Dion, Robert Ross, engineer, Canadian Exploration Ltd., Salmo.
26. Engen, Christine, housewife, Emerald Mine, Salmo.
27. Engen, Wesley Theadore, driver, Emerald Mine, Salmo.
28. Erne, Anne, housewife, Canadian Exploration, Salmo.
29. Erne, Kelvin James, foreman, Canadian Exploration, Salmo.
30. Gritchen, [sic] Nellie Anne, housewife, Canadian Exploration, Salmo.
31. Gritchen, [sic] Paul, operator, Canadian Exploration, Salmo.
32. Hamilton, David John, helper, 109 Morgan St., Nelson.
33. Hartland, Dorothy Vivien, housewife, Canadian Exploration, Salmo.
34. Hartland, George Henry Thomas, Canadian Exploration, Salmo.
35. Heroux, Joseph Laurent Gregoire, warehouseman, Salmo.
36. Johnson, William James, miner, Canadian Exploration, Salmo.
37. Kilford, Nancy, housewife, Canadian Exploration, Salmo.
38. Kilford, Thomas, motorman, Canadian Exploration, Salmo.
39. Kinakin, Carole Phyllis, housewife, Salmo.
40. Kinakin, Harold W., miner, Salmo.
41. Kinakin, William, miner, Salmo.
42. Kowalyshyn, Adam Myroslaw, teacher, Canadian Exploration Ltd., Salmo.
43. Lawrence, Edward Archibald Pepin, engineer, Salmo.
44. Lawrence, Marie Suzanne Marguerite, housewife, Salmo.
45. Le Fort, Beatrice Marie Anna, housewife, Emerald Mine, Salmo.
46. Le Fort, Servant, Emerald Mine, Salmo.
47. Lloyd, John Cecil, crusher operator, Salmo.
48. Lloyd, Myra Jean, housewife, Salmo.
49. Lund, Leslie Hayes, foreman, Salmo.
50. Lund, Olga Annie Alise, housewife, Salmo.
51. McConnell, Eileen Florence, housewife, Canadian Exploration,, Salmo.
56. McLean, Earl Gilbert, electrician, Salmo.
57. Maddison, Beatrice Helen, housewife, Canadian Exploration, Salmo.
58. Maddison, John Henry, welder, Canadian Exploration, Salmo.
59. Neill, Edith Gwen, housewife, Canadian Exploration Ltd., Salmo.

60 Neill, William John, machinist, Canadian Exploration Ltd., Salmo.
61 Nord, Allan Einar, mechanic, Emerald Mine, Salmo.
62 Nord, Dorothy Sophia, housewife, Emerald Mine, Salmo.
63 O'Connell, Julia Josephine, housewife, Canadian Exploration, Salmo.
64 O'Connell, Patrick John, miner, Canadian Exploration, Salmo.
65 Olson, Evelyn Mathilda, housewife, c/o Canadian Exploration Ltd., Salmo.
66 Olson, Melvin Aldred, chief accountant, Canadian Exploration Ltd., Salmo.
67 Owens, George Dale, secretary, Canadian Exploration, Salmo.
68 Pearson, Alison Silvey, housewife, c/o Canadian Exploration, Salmo.
69 Penner, Anthony, mechanic, c/o Canadian Exploration, Salmo.
70 Penner, Margaret Ann, housewife, Salmo.
71 Reyden, Cornelius William, tram foreman, Salmo.
72 Reyden, Vivian Alice, housewife.
71 Robinson, Bernice Elizabeth, housewife, Salmo.
74 Robinson, John William, mining engineer, Canadian Exploration, Salmo.
75 Rose, Gladys Myrtle, teacher, Canadian Exploration, Salmo.
76 Rowe, Lillian Elsie, housewife, Canadian Exploration, Salmo.
77 Rowe, Raymond Clifford, shift boss, Canadian Exploration, Salmo.
78 Rudychuk, William, miner, Salmo.
79 Shelrud, Carl Maurice, miner, Canadian Exploration, Salmo.
80 Shelrud, Evelyn Margaret, housewife, Canadian Exploration, Salmo.
81 Smith, Douglas Reid, accountant, Jersey Mine, Salmo.
82 Smith, Enid, Alfreda, housewife, Canadian Exploration, Salmo.
83 Smith, Olga Emelie, stenographer, Salmo.
84 Smith, Tom Stephens, geologist, c/o Canadian Exploration Ltd., Salmo.
85 Stard, Ronald, surveyor, Canadian Exploration Ltd., Salmo.
86 Stard, Susanne Judith, accounting clerk, Canadian Exploration Ltd., Salmo.
87 Stenzel, Mary Ellen, housewife, Emerald Mine, Salmo.
88 Stevens, Alfred Ralph, machinist, Salmo.
89 Stevens, Mildred, housewife, Salmo.
90 Summers, Alastair Hamilton, miner, Canadian Exploration, Salmo.
91 Summers, Joyce Gwendolyn, housewife, Canadian Exploration Ltd., Salmo.
92 Sztyler, Janusz, miner, Salmo.
93 Takenaka, Tsuneo, technologist, Canadian Exploration, Salmo.
94 Thomson, William Wark, assayer, Canadian Exploration Ltd., Salmo.
95 Van Staalduinen, Clasina, housewife, Canadian Exploration Ltd., Salmo.
96 Van Staalduinen, Matthys, accounting clerk, Salmo.
97 Van Staalduinen, Susanna, housewife, Salmo.
98 Van Staalduinen, Willem, first aid, Canadian Exploration Ltd., Salmo.
99 Verigin, Nancy, housewife, Canadian Exploration, Salmo.
100 Verigin Nick Peter, truck driver, Canadian Exploration, Salmo.
101 Wainwright, Harold, warehouseman, Salmo
102 Weber, Olive Melba, housewife, Canadian Exploration, Salmo.
103 Weber, Robert Gerald, engineer, Salmo.
104 Wilson Berte, assayer, Salmo.
105 Wilson, Blodwen Verna, housewife, Canadian Exploration, Salmo.
106 Wilson, Elsa Christina, housewife, Canadian Exploration Ltd., Salmo.
107 Wilson, Floyd Douglas, security, Canadian Exploration, Salmo.

Certified to be a correct copy of the revised list of voters of the NELSON-CRESTON Electoral District, this 13th day of August, 1966.

G. L. BRODIE, Registrar of Voters., VICTORIA, BC.

Reprinted from a photocopy of the original, courtesy of Mel Olson basis. Ore was retrieved from the reopened Dodger workings, and two 125-horsepower ventilation fans were installed at the portal.

The company's intent in closing the lead-zinc operation was rather than bring on a lot of new people to expand the operations and mine both kinds of metals simultaneously, they would switch over to tungsten and temporarily close the lead-zinc operation until the tungsten was mined out, then return to lead-zinc production. As it turned out, this did not happen, for when the time came to switch back to lead-zinc, changes in the political climate had rendered that ore uneconomical to mine.

A six-foot diameter borehole, 675 feet long, was drilled in 29 working days from surface to the Invincible decline to provide ventilation and a second adit. A prefabricated manway was installed in the borehole in a further ten days. Ore mined during the year averaged 0.60% tungstic oxide, and the mill operated at 500 tons per day, but its capacity was increased to 600 tons by year-end.

The East Dodger was discovered in late 1971, but by May of 1972, the development work in it was complete and the mine was already producing ore.

During that year, all the ore came from the Invincible and the reopened East Dodger. Production was about 500 tons per day. Mining was done with jacklegs and scraper slushing to draw-points, from where trackless equipment transported it to ore-passes on surface. The ore was very irregular and hence difficult to mine.

The extraction of tungsten was an involved and complicated process—one that evolved and was slowly improved on over the years that Canex mined tungsten. The initial phase was to grind the ore to a medium grind in the ball mill, and then feed it over a bank of shaker tables that separated the heavy scheelite from the lighter waste minerals, producing a very high-grade concentrate.

A small, but important portion of the scheelite was not removed in the first phase. In the earlier years this product had been shipped to a refinery in New York for final treatment before it could be marketed. Later the company was able develop a process whereby they could do this themselves.

The first step in this second phase was to heat the material to 55°F (12.5°C) with steam, after which it underwent a flotation process to concentrate the scheelite. This process included treatment with exotic agents such as vegetable oil and a product made from the pulverized bark of Quebracho trees from South America. The resulting scheelite concentrates still contained unwanted minerals including calcites, which had to be removed by hydrochloric acid leaching, after which concentrates still had to be pressure-filtered, producing an end product that was about 67% tungsten (WO_3). This final product was then roasted to remove the sulphides missed in the earlier processes.

The products produced by each of these two phases were then blended resulting in a final product yielding over 80% tungstic oxide (WO_3) which was sold to steel mills in Japan.

From October 1970 to September 1973, the Invincible mine produced 282,779 tons of ore, averaging 0.65% WO_3 (tungstic trioxide), and the East Dodger mine produced 225,094 tons averaging 0.54% WO_3.

Having exhausted the economical tungsten ore, Canex closed the mine in late 1973. Ed Lawrence says that the marketing people at Placer Development had predicted that the price of tungsten would remain stable at about $45 per unit (20 lbs.), but no sooner had the

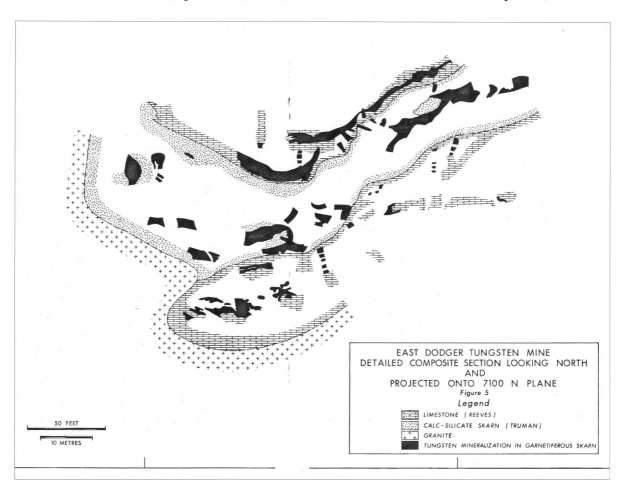

Figure II. 2. Mining in British Columbia Journals, 1973.

Invincible gone into production, than the price plummeted to $24. The company therefore stockpiled 50,000 units and was eventually able to sell this ore at $26 per unit. It seems that the Canex production was enough to upset the global supply-demand balance, for shortly after Canex closed, the price began to recover and reached $125/unit within three years.

■ ■ ■

Mining is a risky business. It requires a lot of capital to prove up enough ore to justify the construction of a mining plant. Added to this is the fact that metal prices can fluctuate wildly, and what is economical ore today may not be so a few months from now. Another problem is one of scale—deciding on the size of the plant to build. The larger the operation, the lower the cost of production generally is. In order to decide on the size of the operation, management must

Table II.2: Summary of Jersey Lead-Zinc Production Figures extracted from the BC Minister of Mines Annual Reports

Year	Tons Milled	Ore Source		Average Crew Size		Development Work (feet)			
		Trackless Mine	Tracked/ Other	Under-ground	Surface	16x16 Drifts	Sub Drifts/ Crosscuts	Raises	Diamond Drilling
1959	325,564	75%	25%	95	59				
1960	364,424	70%	30%	85	75	1,438	7,265	1,439	
1961	374,032	65%	35%	70	87	999	6,227	460	
1962	384,894	83%	17%	99	55	1,282	5,832	608	2,497
1963	368,673	95%	5%	99	68	206	6,776	698	??
1964	407,062	100%		119	68	?	10,363	724	17,560
1965	377,124	100%		86	59	?	9,058	661	24,630
1966	417,440	100%		94	131		8,756	599	13,039
1967	493,029	100%		122	101		12,197	889	16,370
1968	506,220	100%					9,600	759	15,842

have a reasonable estimate of the amount of mineable ore available. This requires a large amount of costly drilling beforehand. Just how sporadic and discontinuous the ore in the East Dodger was, for example, can be seen in the vertical section on figure II.2.

Ed Lawrence says that opening and mining the Invincible would have turned out to be a serious mistake had it not been for the accidental discovery of the East Dodger ore body. The up-front costs of developing the Invincible ore body would never have been recovered because of the collapsing tungsten prices. Meanwhile, the company had been drilling the area east of the Dodger/Jersey area in the hope of finding an extension of the lead-zinc ore, but had serendipitously intersected the East Dodger tungsten instead. As it was, the East Dodger ore combined with that from the Invincible enabled the company to recover most of those development costs.

Ed thinks that the area still has the potential to produce much more ore. He also says that mining as a way of life has been good for a lot of people, not least of all because the work ethic it has fostered.

For two-and-a-half decades, modernizing the mines, switching production from tungsten to lead and zinc and back when the market situation required it, building an exemplary workforce, and having become a leader in training miners, Canex had balanced the risks and had managed to run a profitable business. But ultimately the company had to yield to political change: when in 1972 a provincial government opposed to mining was elected and planned restrictive legislation, all prospects of a promising future were shattered. After the closure of the tungsten mines Canex had intended to revert to mining low-grade lead and zinc, but the new laws governing the mining industry made the prospect uneconomic, and the Emerald was closed down completely. Ritchie Bros., Auctioneers sold off the assets and spent most of 1974 dismantling and shipping them out. Reclamation work was completed, including seeding of the tailings ponds.

Sultan Minerals Inc. acquired the property in 1996 and recently started a new exploration program, which has raised hope that mining in the area will be revived. Metal prices are higher than they have been for a long time and may well continue that way because of rapid industrialization in both China and India. The long-term outlook for tungsten prices appears especially good.

No one can be certain that the Emerald mine will reopen. But in addition to new, lower grade tungsten ore already discovered, there is tantalizing evidence of a huge molybdenum ore body lurking below the old workings. A lot more drilling will be required to prove it up, but with the stubborn optimism of people like those at Sultan Minerals Inc., which presently owns the property, don't bet your money against it.

PART III

"Being There"

INTERVIEWS WITH FORMER
CANEX EMPLOYEES AND RESIDENTS

My own connection to the Emerald Mine was brief—I worked there during the summer of 1952 as a miner in the Dodger 4400 tunnel, usually referred to as the Dodger 44. I had previously worked in four mines and one tunnel, beginning with the Sheep Creek Company's Paradise Mine near Invermere in 1949. When my partner and I left the Emerald and our paycheques were not immediately forthcoming, my partner grabbed the clerk by his tie, hauled his head out through the wicket, and threatened violence if we were not paid immediately. They paid us. They also blackballed us—we could never work there again.

I did go on to work at Remac for seven years, as well as the HB mine for a summer, so the Salmo–Nelson area has always been special for me. This was perhaps the main reason for visiting Salmo during the "Homecoming Reunion" celebrations in August 2006, the town's 60th Anniversary.

I combined this visit to the West Kootenays with a tour to promote my memoirs, which I had published a year earlier. Ed Lawrence, the last mine manager at Canex, was now in charge of new exploration work there. After reading my book, he urged me to write a history of the Emerald. The Salmo Historical Society liked the idea and found money for the project, and Ed's employer, Sultan Minerals Inc., generously supported the project as well.

I began my research by downloading all of the BC Minister of Mines reports for the years during which the mine had operated—probably between 25,000 and 30,000 pages. I also started looking for people who had worked at the mine. Many of the ones I had known were no longer alive, but I did find a few old acquaintances. They in turn gave me names of others, who did the same. It took a while before it dawned on me that I should not only be interviewing the men who had worked at the mine, but also the wives and children who had lived there. I made up a list and kept *expanding it. Each time I interviewed someone, I let them study my list. This almost always led to some new names, and the list grew to over 100 names, of which I have interviewed 90 individuals. There were four people on my list who died before I could see them, so that lent some urgency to my work, as did my own age—I wanted to complete the project while my health was still great.*

Almost all interviews were taped. Except for a couple of cases (where I was asked to proceed differently) I stuck closely to the transcribed texts, rendering the narratives in the first person. I also left miners' jargon and occasionally rough language in, because I feel this makes the personal accounts lively, colourful, and genuine. In this way the stories really bring to life the spirit of the old mining community. Where the individuals talk about their work, we get a glimpse of what that work was like almost fifty yeas ago—the tough conditions, where safety often came second, the competitiveness (there is something distinctly "macho" about a contract miner's work habits and output), but also the strong bond between partners, which is put to the test in adverse situations. Nowhere is this more evident than in the story of the Gil Mosses rescue, when the colleagues of the injured man just wouldn't give up and brought him out of a seemingly impossible situation alive.

While doing the interviews I met a lot of different people. A few were obviously "doers" and not talkers, and what I took home from meeting with them was brief. One former miner, although willing to see me, said, "What can I tell you? I got up in the morning, went to work, worked all day, went home, had a beer, had supper, went to bed, and the next day I would do exactly the same." However, most people I talked to had more to tell me than can possibly be included in this book—I had to make some cuts. Sometimes I left in what the interviewees told me of their lives outside the Emerald, because those stories round out the personal experiences and are what makes up the history of this country; I therefore found them valuable. There are some overlaps, when two people related the same incident. I deliberately left those in, as I feel the different "voices" justify the duplication.

I struggled with the problem of the order in which to present the interviews and in the end decided to present the interviews in roughly the chronological order in which people started to work or live at the mine. Family members appear next to each other, and the last interview is the one with a teacher, who never actually lived at the mine.

The whole project of interviewing former Emerald employees has been an enormously interesting and gratifying adventure for me, for I met so many engaging people I should not otherwise have known.

KEN HENDERSON

When I interviewed Mr. Henderson, he was about 90 years of age, but still living by himself in a large house he had built in Salmo ten years earlier for himself and his wife. His wife died five years ago, and he has lived by himself since then. I was impressed by the ease with which he negotiated the stairs when he showed me around his house. However, shortly after I met him he moved into a seniors' complex at Castlegar.

Ken was at the Emerald while the War Metals Corporation owned the property. He then worked there for Canex as an accountant from 1947 to 1952, before going to work for his father-in-law, Frank Rotter. Here is what Ken told me:

I joined the army in 1939, but was mustered out in Halifax because I had only one good eye and could do only office-work.

I came to Trail from the Con mine at Yellowknife and worked there briefly for CM&S, who then sent me up to Salmo. In Trail they had tossed a coin to see whether to send me to the Pinchie Lake mine or to the Emerald. The Wartime Metals Corp. was operating the mine but asked Cominco to supply the personnel to run it because they were so short of tungsten for the war. Mason was the manager for a while, but Stan Gray who was in charge of all outside mines and his assistant from Trail ran it. The trouble with CM&S was that they were base metal people and didn't know a heck of a lot about tungsten. What happened was they had lead and zinc in the tungsten concentrates and couldn't separate them. They had about 500 fifty-gallon barrels full of tungsten that had lead in it which they couldn't get out. So they put a big magnet above the belt to pull the lead, but it pulled the tungsten with it. They had a tailings pond across the road from the mill, and they assayed higher than the heads *[concentrates]*. At about that time, the Chinese were again supplying tungsten, so they no longer needed it so badly. CM&S was out of its depth trying to mine tungsten.

Ken Henderson, formerly of Salmo, but recently moved to a seniors' facility in Castlegar. 2007.

Colonel Perry had been the accountant up at the Reno mine before coming to Canex, where he was the chief accountant in charge of the tungsten operation. When they opened up a separate office for the lead-zinc, he sent me there. I stayed for only a couple of years and then quit to work in the lumber business. While I was at the mine, the colonel lived next door to me in the houses behind the main office. Jim Magee lived on the other side of me. I had known him from Yellowknife as a miner. After he took his university training, he came back as a mining engineer.

One of the men on the company's advisory board was mixed up with the Panabode company which manufactured these knock-together houses. He got the company to put forty or fifty units in at the Emerald. When I got married, I got one of them. It was the poorest house I've ever lived in.

We had a miner living next to us who had brought back a Scottish war bride. They had these two young kids playing in a sand box beside the house and I see this bear going through the garbage, not ten feet from them. So I carried my rifle with me to the office and shot the bear. I told the guys to take it away, but pretty soon there's another bear down there. After I shot the third one, the guys on the truck said, "We're not taking anymore bears away for you." I asked my wife, "What'll we do with it?" She said, "Let's eat it. Bear meat's good, and it's a young one." So I carved it up and we invited the young man and his war-bride over for dinner. His wife said, "That's the best meat I've ever tasted. What's the name of your butcher? I want to get my meat from him." When I told her she was eating bear meat, she threw up.

I left the mine and went to work for Rotter *[his father-in-law]*, who bought a house for us in Salmo. At first I handled the shipping and the accounting, but Rotter wanted me to learn to run the mill, so I studied for and got my scaling licence, lumber grading, and first aid and became the foreman. Frank was death on unions. He paid his men as little as possible and was a hard guy to work for. When unions came in and started to tell him how to run his business, he sold it. He couldn't take it. The union guy who was the head of all the unions came up on his motorcycle and started to tell Frank. Frank hauled off and knocked his teeth out and then he locked himself in the cookhouse. This guy came out there calling him everything under the sun to get him to come out and fight. Frank laid charges against him because his pal (Johns) was the judge in Salmo; Johns fined the union guy.

ANN HENDERSON

I interviewed Ann Henderson, Ken Henderson's daughter, in the village office one afternoon in the fall of 2007. She is the first female mayor of Salmo and is still in her first term. Another first for Ann is being available to the townspeople by spending every Monday in the village office.

I lived at the mine for a couple of years when I was six to eight years of age. Many years later, I returned and worked as Ed Lawrence's secretary for over a year during the 1969 to 1970 period.

Ann Henderson.

The Panabode house we were in had cracks between the logs which we couldn't chink for a year until the logs had finished shrinking and settling. Mother sometimes wore a fur coat in the house, and I remember the snow coming in through the cracks and settling on the newspaper Dad was reading.

One unforgettable memory was my one-year-old-brother playing in a sand box beside the house. Mother looked out and let out a ghastly scream. A bear was sitting in the box playing with my brother, but the neighbours who came running in response to mother's screams quickly frightened it away. I also recall the staff lodge burning down as well as a Panabode house.

When I started school, I had to ride into Salmo on the bus that took the miners from graveyard shift into town. I came back home on the bus carrying the miners who were on the afternoon shift. By the time when I was in grade two, they had built the Harold Lakes School, and I attended it for that grade. We once had a mean teacher who made the seven-year-olds clean the school toilets. However, Mother quickly put a stop to that practice.

I worked for the oil and gas companies in Calgary as a certified paralegal for a number of years; then I married Jim Bakken in 1989 and came back to Salmo. Jim, who in addition to his regular duties at the mine, built mine ladders on contract outside of his normal working hours.

STAN HUNTER

I interviewed Stan Hunter in his home on Churchill Street in Vancouver. He impressed me by being a man who had a good word for everyone he had worked with. He is also an entertaining writer. A number of his articles about the Emerald were published in the Northern Miner, *and some of them are reprinted in this book. He was at the Emerald from 1947 to 1951.*

I was born in Vancouver, but raised on a farm at Hazelton. I took my high school there through correspondence courses and graduated

Stan Hunter, B.Sc. (Eng., Mining), of Vancouver worked at the Emerald in the late forties. 2006.

from UBC as a mining engineer in the spring of 1948. There were only six other men in the graduating class, where we had an Australian professor named Les Crouch—a very competent man who had worked all over the world.

One summer I got a job at the Red Rose tungsten mine near the top of Roche de Boule Mountain west of Hazelton. The whole camp was cabled down to the mountain to keep it from blowing away in the sudden, powerful gusts that frequently swooped down on the camp. The buildings and the portal were connected by a series of snow sheds, which were also tied down.

The Consolidated Mining and Smelting Company, commonly referred to as CM&S and later as Cominco, was developing the mine. The foreman was a man named Jack Zucco—one of Cominco's best. His job for me one week was helping the cook. The following week it was mucking by hand behind the miners in the drift and then a week on the aerial tramway sending the ore down the mountainside to the concentrator. The job did not last long, but it was long enough for me to decide that mining was for me. I learned much of my mining from Jack.

During UBC years, I worked my summers for C. A. Banks of Placer Development. They had an office in the Royal Bank building at Hastings and Granville in Vancouver, and, in the company of H. L. Batten, Ron Murphy or Jerry Christie, I travelled all over the province assessing potential mines. They wanted to explore everything. We first went to Stewart, an out-of-the-way place north of Prince Rupert. Next we went to Salmo in the Kootenays, a place I had never heard of. Old Banks had us patrol the whole province looking for prospects. In Batten he had hired a man who had an open mind. He knew the province and he knew mining; Jerry Christie likewise, he had been all over the province and had surveyed for the railways as well.

In 1947 Christie called me and said, "Stan, the old man has taken an option on the Emerald, which is up for sale by the Wartime Metals Corporation. He bought it for peanuts and we want you to join us at the mine this summer." At the mine they had the advantage of hiring old experienced Depression era miners who knew how to work and how to mine. There I worked with old-time miners like Jack Anderberg, who was a heavy taskmaster for me. He was forever on my neck! He wanted things done right. If I know mining, it is because of working with men like Anderberg, Sig Nelson, and people like them.

One early job I got was surveying the tramline from the mine down to the mill. Can you imagine me, a student, surveying the whole tramline? Christie, who was the chief engineer, next had me study the claims map, then told me, "We want to expand our holdings. I want you to start at the Jersey and stake the rest of the mountainside." One fraction I found is named the "Stan Hunter Fraction," but of course I did not own it. I also did a lot of surveying underground, where I worked closely with Jack Anderberg and Sig Nelson.

After graduating, I went to work for Placer again and was sent to the Emerald mine. We had some tough times, and one month we would be in the doldrums when we couldn't market the tungsten, then the next month the prices would improve, and we were all happy again. They hired a new manager, Rollie Legg, from Copper Mountain, who brought a fresh perspective to the mine. The mill continued to pose the most problems, for we had trouble meeting the specifications.

Elmer Lonergan, Doug Little, Stan Hunter, and Lawrence Adie. Early 1950s. Courtesy of Marie Adie.

In Salmo I stayed at Ma McKeon's for a while *[she was the local bootlegger]*. I thought it was a great little village. The beer parlours on Saturday nights came alive with all the miners. The town had ball games in the summer and the hockey games in winter. It had a real good crowd.

We had started to mine the Jersey while I was still there. The diamond drilling had indicated a large lead-zinc ore body and we had begun the first drifts into it before I left there. The trackless mining began later.

I shortly moved on, and my wife followed me everywhere I went. Late in my career I wound up opening a mining school at Rossland, BC. Premier W. A. C. Bennett summoned me to Victoria and said, "Look, Stan, we have 50 mines in the province, but we are short of miners. I'm thinking of opening a mining school in Rossland and training young miners. Would you be interested?" I spent eight years in Rossland and turned out 2,000 young miners for the industry from 1971 to 1978. We started with the Red Mountain mine, then moved to Cominco's little lead-zinc show nearby.

Bennett would call me up at Christmas and say, "Now listen, you little old son of a bitch. You're spending too much money. Cut back on the costs. Turn out more miners at less cost." That was his annual Christmas greeting.

I think the school was quite successful. The kids all got jobs and did well. We turned out the first women into the mining industry. One girl who went to work at Endako became the first female mill foreman in Canada. I couldn't believe it. She also became the mayor of Fraser

Arvid (Bert) Lundeberg; he probably stayed at Canex longer than anyone else. 2007 photo.

Ray Rowe and Bert Lundeberg poring over a plan. Courtesy of Sultan Minerals Inc.

Lake. I used to drop in to see her when I was passing through after I became Mine Inspector.

I think the Emerald was a great training ground for young engineers, geologists, and metallurgists. At the Emerald during Banks's time, if you had a criticism to make, you made it. If you had an observation to make, you made it. However, I never did come back to the Emerald. I enjoyed mining and especially the Emerald. It was my first solid venture into mining and it was a good one.

ARVID (BERT) LUNDEBERG

Bert is still going strong and is now about 90 years of age. Bert's account of his many years at Canex is short—he must have been a doer, not a talker. Owen Bradley told me that Bert had been in one of the first landings at Normandy during World War II. His one complaint was that as he was scrambling down the rope nets on the sides of the landing craft, the men following him were continually stepping on his fingers. Bert worked at Canex first as a miner and then as a shift boss. Except for one short period, he was there from 1947 to 1973, when the mine closed.

I worked at the Reno mine for a year and a half, beginning in 1937, after which I worked at the Arlington Relief mine near Ymir before joining the army in 1940. In the army I trained at Fort Garry on 25-pound Howitzers and did some instructing in Camp Shiloh. We were in Britain for several years before going to France 14 days after D-Day. I was in Holland when the war ended and stayed there for some time. I eventually came back to Canada and was discharged in January of 1946. Meanwhile, I had met my Scottish wife-to-be in England. Unlike many others, I heeded the orders not to marry without permission. Because of this, I had to pay for my wife's plane ticket to Canada and we also missed out on "wife allowance payments." We were married on Jan. 18, 1947, and recently celebrated our 60th anniversary.

On my return to the Kootenays, I went to work at the Emerald Mine but left in 1948 to work for Northern Construction in the Kimberley tunnel for eight months. I returned to Salmo and to the Emerald, where I worked as a miner until 1963 and then as a shift boss until it closed in 1973. I was on duty the evening, but not in the area where Gil Mosses was being rescued from the chute in which he was trapped.

I first worked in the glory hole [open pit] and then in the Jersey. We benched the stope down until the floor was 90 feet below the back [roof]. The rock in the mine was very stable, for twenty years later when we scaled that same back from a Giraffe, we found only a couple of small pieces of loose rock.

Scaling from a Giraffe. Courtesy of Sultan Minerals Inc.

My wife and I both knew Harold Lakes well, and I remember that he kept a black cat to which he regularly fed pork hearts. In fact, he had a small fridge for the sole purpose of storing these goodies for his cat. We also knew Clive Ball, the geologist from Australia. He had been a transport pilot and had flown the "hump" during World War II. This was a difficult and dangerous supply route from India, over the Himalayas to China, flown with unarmed freighter planes.

THE HARTLAND FAMILY

There were two Hartland brothers who worked at Canex for many years. George, the older brother, was an equipment operator, and his younger brother Denis became a mill man there. George is deceased, but his widow, Dot, an English war bride, now lives in Whitehorse. I met George and Dot once, but did not know them well. They had two sons, whose interviews follow after their uncle Denis Hartland's interview.

I have known Denis since 1955. Until I left the Salmo area in 1962, my wife and I used to visit and play cards with Denis and his wife Shirley while they lived at the Canex lead-zinc concentrator. I had first known Shirley as a schoolgirl in 1952, when her father was the electrician at Remac where I then worked. They are divorced; Denis has lived in Williams Lake for many years and Shirley lives at Comox. Both George and Denis were at the mine for most of the years that Canex operated it.

George and Dorothy Hartland in mid-fifties. Courtesy of Lewis Hartland.

DENIS HARTLAND

Denis Hartland inside his log home north of Williams Lake in late 2006.

I was born in the Edgewood–Needles area on January 11, 1929, orphaned at age six, and looked after by my two older sisters before going into foster homes. I was related to Bob Hallbauer (former president of Tech-Cominco), who once saved me from drowning in the Whatshan River near Needles, BC. *[How appropriate that Denis would years later save two men from drowning in Osoyoos Lake and one man from drowning near Erie west of Salmo.]*

My first job at the Emerald was as "flunky," where I started on June 17, 1947. However, I quit the following year and worked eight months as a labourer in the hydro tunnel near Needles for Miners Western, a tunnel contractor. I next worked in Alberta before returning to BC and working as a longshoreman in Vancouver. In June 1950 I returned to work as the second cook for Ace Bailey at Canex for $90 per month including free room and board.

Ace was away on days off. Before leaving he had given the guys permission to use his radio to listen to the World Series ball games. When he returned, he was drunk and his radio was playing music. He immediately

Denis Hartland. Courtesy of Denis Hartland.

Denis Hartland in Vancouver, July 1949. Courtesy of Lewis Hartland.

demanded to know who was using his radio. A flunky told Ace that he himself had given the okay before leaving. Ace grabbed a big butcher knife and chased the flunky, who ran around the table trying to stay out of the enraged cook's reach. As Ace ran past me, I stuck out my foot and tripped Ace, who went sprawling. I then led a walkout of the kitchen staff, demanding that the company get rid of Bailey. Canex fired me instead and blackballed me as well, but Marshall McDiarmid, the foreman at the concentrator, promptly hired me anyway.

The ladies of the Emerald Mill on March 30, 1953. Back row, left to right: Unknown, Edith Black, Gladys Erickson, Unknown, Mary Berryman, ? Swafford. Front row: Edith Dawson, Margery Smithson, Betty Hobbert. Courtesy of Ron Erickson.

Things must sometimes have been tough during the early years, for I remember the manager coming into the cookhouse one payday and saying to the staff, "Due to financial problems you won't be getting paid for a few days. Please bear with us." I also remember going down to the dry and washing the mechanics' wiping rags in order to save a little money.

The tramline was still operating in 1949, but its wooden towers kept falling down in high winds. So an American company rebuilt it. They installed steel towers on concrete foundations but when the mill was expanded in the early fifties, the tramline could not handle the volume, so Canex put in a conveyor system to replace it.

Shirley Williams and I were married at Coeur d'Alene in May of 1955, and I continued working in the mill, first as an operator and later as a foreman. During the labour strike in 1959, I worked for Intermountain Construction at Bethlehem Copper for a couple of months but returned and stayed at Canex until January 1973, when I took time off after my marriage broke up. I had intended to quit, but Ed Lawrence, the mine manager, persuaded me to take off as much time as I needed. When I returned the mine was closing, so after deciding against a job at Craigmont, I went to the Gibraltar mine near Williams Lake where Jim O'Rourke gave me a choice of jobs. I chose hourly work, but before long I was the foreman in charge of the tailings pond.

LEWIS HARTLAND

I finally got to meet Lewis (and his interpreter) in the fall of 2007, when they came down from Whitehorse for a doctor's appointment.

Lewis became deaf from meningitis at age eight months. For some people, such a handicap would have been an excuse to give up on life. For others, like

Lewis Hartland from Whitehorse, 2007. He established the "Canadian Deaf Theatre" in 1989.

Lewis's friend and interpreter Michele, 2007.

Glen and Lewis Hartland. Courtesy of Ernie Stenzel.

Lewis, it became a goad that drove him on to achievements he might not otherwise have reached. He went to a school for the deaf in Vancouver from 1960 to 1971.

In 1970, Lewis took instruction from a professional mime instructor at Simon Fraser University. In 1977 he trained at the Canadian Mime Theatre School in Niagara on the Lakes. He followed up with actors' labs and other training. From 1977 to 1989 he presented solo mime performances throughout Canada and the United States. He also founded Theatres of the Deaf in Ontario as well as in other places. He finally helped found the Canadian Deaf Theatre (CDT) in 1989—the only professional Anglophone deaf theatre in Canada. It presented its first production, starring Lewis, on January 10 the following year.

Lewis now lives in Whitehorse, where through his efforts City Council now provides closed-captioning for its council meeting broadcasts—the first Canadian municipality to do so. Since 1997 he has owned and is president of "Visual Interpreter Services of Yukon." He is now starting a new business as well—providing the area with a variety of smoked meats that includes salmon and a variety of jerky made from beef as well as from wild game.

Here is Lewis Hartland's account of his time at the Emerald in his own words:

I remember kids throwing rocks at me when I was between three and eight years old. They often beat me up too. There were two brothers who threw a huge rock at me when I was three years old. It hit me in the face and there was blood everywhere. A neighbouring man came and picked me up and carried me to my house. My mother and the neighbour lady then took me to the clinic in Salmo.

During my childhood I took a lot of beatings, but I have managed to stay positive. When kids stole my toys I would chase them to get them back. I played with girls a lot and always got along with them, but the boys really teased and beat me because I was deaf. They thought I was dumb, because I couldn't talk. I remember a girl and I running for our lives when the boys chased us with wooden sticks. My older brother Glenn once built some forts for me and the same kids who used to beat me destroyed those forts as well.

I always had communication barriers. From age seven until I was a teenager, I used a stick to write notes in the gravel. That's how I communicated with the other children. I supplemented this by improvising with my own gestures as well. I liked hanging around near Ernie Stenzel, for he and the other older boys were always good to me and respected my deafness, as did the older girls too.

When I was three years old, I started to climb trees so my mother sometimes had to go to the school and get Colin to come home and get me down. I loved to climb trees because it was quiet, and me being deaf, I could see everything, being on top of the tree. My mom still remembers that. I guess it was fine with me. There was no communication with the other kids—they were all in school.

I did a lot of swimming at the Vancouver school and went to provincial meets as well. I also won a lot of ribbons at the Emerald Mine swim meets during summers.

From age 5 to 17, I went to school in Vancouver and came home at Christmas, Easter, and for the summer holidays. By the time I graduated my parents had moved to Creston, and the mine shut down two years later.

Being born without hearing led me to develop my other senses more fully. Patricia Wlasiuk remembers me being aware of approaching

Glen Hartland, Bob Adolphson, Ernie Stenzel, and Bob Krasnikoff. Courtesy of Ernie Stenzel.

Lewis with older brother Glenn on skates in 1960. Courtesy of Lewis Hartland.

traffic before they could hear it. I think that I just used my eyes more effectively than most hearing guys who depended on their hearing as well as their eyes. We live in two different worlds.

When I was a little boy, I was a fishing freak. In the summer time, I constantly bugged my father, "Dad, please let's go fishing." Dad would sigh, "Be patient, son." But we'd eventually go. We did a lot of camping too. My mother, father, and Glenn too would come camping with us. My brother Colin didn't come very much, for he joined the army and went off to Chilliwack and then on to Germany."

Lost Creek is where I usually fished. It meandered between Iron Mountain and the adjacent one to the south. Fishing became a tradition with us. My older brothers Colin, Glenn, their buddies and peers all fished there as well. During the summer my young friends and I often walked down the mountain to Lost Creek from the mine to fish for char. We usually stayed there from morning to late afternoon before trudging back up the mountain to our homes, sometimes lugging lots of fish.

On one occasion my friend Brian and I biked down mountain with our rods to Lost Creek. We hid our bikes amongst the trees, and then fished from morning until afternoon. When I went back to get my bike, its seat was all chewed up. I had won this particular bicycle as a prize from a cereal company. I had participated in a cartoon drawing contest, and to my surprise, they picked me as the winner. The bicycle had the banana shaped seat and long, high, and wide handle bars. I first thought that a bear had attacked it, but it was a porcupine. Like many animals, they like salt, and it could smell my sweat on the banana seat and ripped it apart to get at it. We walked back up the mountain with our bikes, where I taped the seat back together again as best I could.

During the summer time there were always a lot of black bears around. I remember one bear hanging around. One night my friend and I chased it up a tree. Two dogs kept barking at it and I got the hose and shot water up the tree and soaked the bear, but it stayed there refusing to give up and come down. I was the one who finally gave up and went home. Those are great memories.

COLIN HARTLAND

Colin is the son of the late George Hartland and is Denis Hartland's nephew. He lived at the mine from 1954 to 1963. I met Colin when he visited in the Lower Mainland in early 2007.

I started grade one in Salmo and attended there until we got a house at the mine in 1954, at which time we moved into a matchbox duplex in 1954 when I was six.

Father worked underground, but not for long. He went on to road maintenance equipment and took care of all the mine roads, including the interior ones and the ones down to Highway 6 all the years we lived there.

We had the only heated swimming pool in the region. I had no idea of how to swim when they built it, but I took lessons that first summer, and once I learned how, you couldn't keep me out of that pool. The pool was heated with the cooling waters from the mine compressors, and people came from all over to swim in it. We swam there even on cold days or when it was raining, but we often had trouble getting a lifeguard to sit out there to watch us when that happened.

Colin Hartland, son of the late George Hartland and nephew of Denis Hartland, 2007.

After swimming, we would hike down the mountain to Lost Creek, do some fishing and hike back home with a bunch of fish for dinner. Today people would think twice about that. It was not only a strenuous walk, but there were a lot of bears around. Of course we paid them as little heed as they did us. I don't remember a boy on that mountain that didn't fish. We never had much money, but nor did anyone else.

When fall came we went grouse and deer hunting. Once we were twelve or thirteen we could get a junior hunting licence. We were then allowed to hunt as long as we were with an adult who had a regular licence.

We used to have a company as well as a union picnic each summer. In addition to children's sports, there would be mining competitions

The swimming pool under construction, mid-1950s. Courtesy of Colin Brown.

Even the grown-ups like the wading end of the pool. Courtesy of Cliff Rennie.

Swim Race. Courtesy of Carol McLean, née Anderson.

JEWEL OF THE KOOTENAYS / 75

and logging events. It was fun to go and watch some of those big tough guys go at it. The mine also had a mine rescue team, for which I sometimes played the injured miner.

As children we were well taken care of. There were always sports; softball, baseball, skating, hockey, and curling. We also had Cubs, Brownies, and Scouts, and I was very much into the scouting movement. Everything was thought about. You did not have to get involved, but of course you did, because all your friends did. I have a picture of a group of us scouts, but I might no longer be able to name everyone in it. I used to have photos looking out over the top of the clouds, which appeared like a lake.

Bears were commonplace in the townsite. They were forever tipping over the garbage cans, and we would sometimes swat them with a broom to shoo them off the porch.

I remember lots of snow and cold during the winter. We did not have warm boots to wear like we do now. The last house we lived in was terrible to keep the snow off. We cooked with an electric stove and heated with an oil stove, but it was not good at keeping the house warm. Sometimes it would go "poof" and die, so we also had portable electric heaters to help keep us warm.

George Hartland in Vancouver during mid-fifties. Courtesy of Lewis Hartland.

I left the mine when I was sixteen and went into the Armed Forces. I spent the next three years with it in Germany, and four in Chilliwack. After leaving the service around 1970, I worked in the mill at Endako, but did not stay because they were slow in getting me a house, so I came back to Salmo and worked at the mine and stayed with my parents for a short while in '71 to '72. They were getting the tungsten mill ready to go again, and we hauled balls and stuff up for it.

I also looked after the compressors while the company was in the process of switching over to an automated system from the old antique Ingersoll Rand piston compressors. I had to be in constant communication with the shift boss to make sure they had enough air. It seemed they were always crying for more air, for compressed air was the heart of the operation. Without the compressors, the mine wouldn't run. I think we had five units in the compressor house. They were huge and were powered by big electric motors. Water was continuously piped in to cool them, and that water was then used to heat the swimming pool.

I have two brothers, one of whom was born in Nelson while we lived there. He now lives in Whitehorse, as does my mother, who is about 82. My other brother, Glenn worked at Canex for a while driving underground equipment. He now lives in Vernon.

I still like to go back and look around. When you go back to a place like that it awakens fond memories. The houses are all gone, but I can still find where we lived. There are still remnants of the garden my dad made. The sad part for me is not being able to go back and find much of what I remember from my childhood there. I will always have great memories of growing up in the bush. The city will never be for me.

Lewis, Colin in uniform, and Glenn, circa 1960. Courtesy of Lewis Hartland.

A FAMILY AFFAIR

The Stard family provided four men and one woman to the Canex workforce. This must surely constitute some kind of a record. Tony Stard, the father, toiled at the mine for twenty-five years, from 1948—shortly after it opened—until 1973, when it closed. His sons and a daughter-in-law worked for the company for significant time periods too. Tony, the eldest, was there from 1963 to 1973, except for a year or two up north. Ron, the second son, worked from 1962, when he graduated from high school, the mine closed and he transferred to Endako. John, the youngest, worked at the mine from 1966, when he graduated, until 1970. Ron's wife Judy worked in the Canex office from 1964, when she married Ron, until 1967, when her daughter Rhonda was born. Tony Stard's son-in-law, Ed Gladu, who also worked at the mine was involved in the Gil Mosses rescue, and received a bravery award. This remarkable family worked an aggregate of over 50 years at Canex.

Tony Stard Sr. lived at Salmo for a while, but then moved his family to a house at the lead-zinc mill site, where he remained until he retired at age 65, which was also about the time the mine closed. He was a respected miner who not only worked hard, but raised a family of four: three sons, Tony, Ron, and John, and a daughter, Vera.

Tony Stard Jr. was not available for an interview, as he was getting ready to leave and was going to be away for several months. He worked as a miner and occasionally as a shift boss all the years he was at Canex. Being the eldest in the family, his wages were probably indispensable to

Mrs. and Tony Stard with daughter-in-law Judy at Mine Rescue competition in Nelson. 1967. Courtesy of John Bishop.

JEWEL OF THE KOOTENAYS / 77

the family, so unlike his younger brothers he was unable to finish high school. His wife, Ada, worked many years in the Canex commissary.

RON STARD

I interviewed Ron Stard, who now lives in Mexico, by telephone in December of 2007. Because he is on Vonage (voice-over-internet protocol), the cost was the same as calling him in BC—five cents a minute.

Tony Stard's son Ron and wife Judy. Late 1960s. Courtesy of John Bishop.

Dale Owen, the mine manager's secretary; Ray Rowe, mine captain; Ron Stard. Circa 1967. Courtesy of John Bishop.

I graduated from the Salmo high school in 1962 and went to work for Curly Colwell in the warehouse at Canex for the summer with intentions of joining the RCMP. However, as soon as I made some money, I bought a car. One thing led to another, and in 1964 I married Judy, the stepdaughter of local hotelier Ben Reid.

At work I became a geological assistant to Jim Bristol for a while, then a surveyor, and before long, the chief surveyor. Following that, I took on the duties of the planning engineer, followed by a position of mine engineer. I next entered the training program to learn how to mine. I followed this up with two years as assistant mine superintendent, and when they opened the Invincible, I was the chief engineer there. I remember Al Sommers and Ed Lawrence as two great guys who did more than their share of training. Sommers would often spend time teaching, both before and after the regular work hours. When the mine closed in 1973, I transferred to the Endako mine near Fraser Lake as one more graduate of the Canex School of Mines. If the Emerald was to reopen, I would be strongly temped to go back and work there without pay.

I spent five years at Endako: first as a shift supervisor, then as the "drill and blast" superintendent. My next job was as drill and blast superintendent for Cardinal River Coal near Edson, Alberta, where I was interviewed for the job by

Dale Owen (formerly of Canex). I stayed with Cardinal River Coal from 1978 to 1990 and rose to the position of the assistant manager in an operation that employed 850 people.

In 1990 I transferred to the Luscar Sterko mine near Hinton, Alberta, where I stayed until I retired at age 53. They had offered me the position of mine manager, but I declined it because I was retiring. After I retired, my wife Judy and I moved to Lake Chapala in Central Mexico, where we still live. There are many Americans and Canadians here, including my brother John.

JUDY STARD

I then spoke with Ron's wife Judy, who told me a little about her life at the mine.

I was born at Didsbury, Alberta, a small place fifty miles north of Calgary. I came to Salmo with my mother when I was in grade 9. When Ben Reid, a local hotelier, married my mother he became my stepfather.

After marrying Ron in 1964, we moved into a house in the Jersey townsite. I went to work in the Canex office, where I was on the switchboard or doing steno work for Mary Stenzel and others. I left before having my first child, Rhonda, in 1967. She is now married, and they had an Esso bulk fuel dealership. Later we also had a son, Rodney. He now works in Public Works in Surrey, BC.

Grace Ash and Mary Stenzel. Courtesy of Sultan Minerals Inc.

JOHN STARD

In a separate call I spoke with John Stard, who lives near his brother in Mexico.

I was born in 1948, about the time that my father began to work for Canex. Ed Lawrence hired me as soon as I finished high school and sent me to work for Al Nord on the conveyor belts, but it was not long before Ed moved me into the engineering

John Stard, surveyor, Ron's younger brother. Mid- to late 1960s. Courtesy of John Bishop.

Ron Stard operating an electrical slusher. Mid 1960s. Courtesy of John Bishop.

office. There, in addition to my other duties, I learned how to survey, a job I kept until leaving the mine in 1970.

After Canex, I went to work for Echo Bay Mines near Port Radium, NWT. It was a small silver mine, where my job was in the engineering department. After four years there, I came back south again and worked two years for Columbia River Mines near Golden, BC. It too was a small silver property in the Purcell Mountains at an elevation of over 6,000 feet. I spent the next two summers at the mine and the winters in the Vancouver office before moving on again.

I next spent eleven years with Sherritt-Gordon, first at the Fox Mine at Lynn Lake, where in addition to my engineering duties they trained me as a shift boss in a mine that produced 6,000 tons per day. The company sent me to a mining technical college for two years. After that they transferred me to the Ruttan Mine at Leaf Rapids as a supervisor, then as a mine captain and finally as mine superintendent in a 6,000-ton-per-day operation. In Lynn Lake I married Sylvia, who was from Kirkland Lake and a nurse.

My next employer was Placer Dome at Timmins, Ontario, where I spent five years as Mine Superintendent Detour Lake.

It may seem like I wanted to see Canada from coast to coast, for my next employer was Royal Oak at Hope Brook in Newfoundland, where I stayed for three years as the superintendent in a 3,000-ton-per-day mine, before travelling north to the Yellowknife to finish my career as the general manager for Miramar, first at the Giant Mine and then at the Con Mine. Eventually I had to preside over shutting down these two old mines that had operated for so many years. I then retired and joined my brother Ron in the Lake Chapala area of central Mexico. Ed Lawrence claims my brother and I as graduates of his Canex School of Mines.

MARIE ADIE

Marie married Lawrence (Joe) Adie, a geologist, and moved up to the mine in 1949. What was it like for a woman to marry a fellow and then move into a company townsite where she likely had no car to drive, and if she did, she probably did not know how to drive anyway? She got to town to buy groceries only twice a month and had neither a telephone nor a television set.

I was born in 1924 in an old house that had been turned into a nursing home on High Street in Nelson, BC. My mother, being 41, thought it better to have me in a special facility rather than at home. The place turned out to be special alright—it was an abortion facility, but Mom had no inkling as to its real nature. A nurse came over and said to her, "You just say the word, dearie, and we will soon have this all over with. You're too old to have a baby."

I was a mistake. I came much later than my brothers. I was 18 months old at my oldest brother's wedding, and my youngest brother was 15 years older than me. Mother said she cried for three months until she decided one day that she had better dry her eyes and get on with it.

My parents came from England around 1900. Dad ran the Nelson Gas Works, across from where was the old smelter was. There they generated gas from coal that came from Fernie. Producing gas from coal was common at one time, and they did it in Vancouver as well, before natural gas became available. The plant had huge furnaces that not only produced gas, but coke as well, a fuel we burned in our heaters on each floor of this huge old house we lived in. Those heaters would get red hot some times.

Before we were married, Joe stayed with us, in my parents' home, for a while. We moved up to the mine in 1949 after being married in Nelson and stayed there until 1954. Harold Lakes was the manager then and his wife Barbara was up there with him. Barbara Lakes was very English, but she was a real frontier person. Her parents had told her that when she moved to Canada, she would have to learn to rough it and just take whatever came. She was great.

When we arrived, there was no proper housing. The manager lived in a tarpaper shack near the bunkhouse until they built a nice manager's residence near the main camp. They built a lot more housing after they opened the Jersey mine.

When I was first married I had to have a clothesline, for there was no such thing as a dryer then. I had a very small portable device but it was inadequate, so I kept

Marie Adie in 2007.

Joe Adie and Elmer Lonergan. Joe (as in Joe-Boy) was a nickname, signifying he was everyone's errand boy. 1950s. Courtesy of Marie Adie.

Marie and Joe Adie. 1950s. Courtesy of Marie Adie.

Marie with baby. Early 1950s. Courtesy of Marie Adie.

Picnic at Rosebud Lake, on the road to Remac, just west of the Nelway Customs. Early 1950s. Courtesy of Marie Adie.

bugging Joe to dig a hole and plant a pole for it. He finally took a pick and went out and started swinging it to dig this hole. He didn't know that the water supply for the camp was there and he put the pick right through the pipe and created a geyser. The camp was without water for nearly a week. Everyone soon knew about Mrs. Adie's clothesline cutting off the water supply.

We were all young and we were all poor at the same time. The Hallbauers didn't have a washing machine. My mother had an old Beatty, which I gave to them because they had a baby. That washer had already done its duty before Joan got it. I had one of the first automatic washing machines at the camp. Everyone was impressed, so I invited the women over for coffee and we all sat and watched a load of laundry go through. It was a great luxury to us at that time.

The children at the mine may not have had many toys, but they had lots to do on the hill. They played together and had the freedom to roam the mountainside. Then they built the swimming pool. My husband and the others all worked on it, but we moved away before it was finished.

We had no babysitters, but we used to help each other out in that regard. We were a close-knit community and took care of each other because we had to. There was just one telephone when I arrived, but the operator was in Salmo, and she shut down the switchboard at 8:00 p.m. If anything happened after that, we were on our own.

We went to Granby for about a year, but then we came back to the Jersey where we lived in a duplex. Those Panabodes were something. Sometimes when it snowed, it blew in through the cracks between the logs.

When Ace Bailey was sober, the food wasn't bad, but when he was drunk, holy cow! One night we had a dance in the community hall, which was also the kitchen, and Barbara Lakes was looking for the ladies' room. She was directed upstairs and told where to find it. She headed up and there was Ace, drunk as a skunk and having just come out of the shower, all he was wearing was a towel.

At one of those dances I went into the kitchen to get some water. We had this big walk-in freezer and I see these legs sticking out from inside of it. It was Billy Gray who had gotten drunk, passed out, and fallen inside the freezer, but fortunately the door would not close or he would have frozen to death.

I used to order the food for the cookhouse, which was always top grades. One of the men, who was an electrician, had a beautiful black Labrador dog. Occasionally we would see this dog trotting between the cookhouse and the office with a huge "black as a cinder" T-bone steak it its mouth. Someone would say, "Oh oh, Ace is drunk again. He burnt a steak." He was some character, but we had a hard time getting and keeping cooks because we were in the bush.

It was a bit scary living up there at first for there was only one road in, and during a hot dry summer, with everything dry as tinder, they sometimes closed that road from 8:00 a.m. to 4:00 p.m. They were concerned about someone throwing out a burning cigarette and starting a fire.

Colonel Perry's wife Elaine loved to garden. When she dug down into the ground in May, she found it was still frozen. Her husband was a great guy. He had osteomyelitis and had only one good leg. He loved to play chess with my husband Lawrence as well as with someone in Vancouver by correspondence. The correspondence game went on for ages. They were much older than us and retired to Abbottsford. He was quite a character.

Elaine Perry and Marie Adie. Early 1950s. The Perry couple were the godparents of Marie's children, Ann and Barry. Courtesy of Marie Adie.

Bob Weber was a really nice fellow. He was killed in a CP Air plane that was blown up in the air near 100 Mile house. His wife, Melba, lives in the area and works in the West Vancouver Seniors' Centre. I was fortunate that Joe wasn't on that flight, for Bob had asked him to come with him, but my husband had something he had to finish up first.

Tony Triggs got up one night when he heard a noise in his back porch. It was a bear. The bear tore down the mountain side. It ran into a telephone pole that was connected to the fire alarm for the whole camp. The thing went off and awoke everyone at one o'clock in the morning.

Management had warned Lieutenant-Governor Banks that the roads to the mine were terrible, but Mrs. Banks insisted on her limousine. That poor chauffeur. Every time he came to a switchback, he had an awful time getting that vehicle squared around for the next leg. It was back and forth, back and forth, several times for each one of them. We had only the north road into the camp in those days. However, they finally made it to the mine.

Even though Barbara had no help, Mrs. Banks insisted on a formal dinner, Government House style, every night, with them dressing accordingly. Johnny Moran, who lived at Thrums, used to bring a movie every Friday night to show in the cookhouse. To get go see it we had to walk from our place down a little path, past a window in the manager's residence, and over to the kitchen. As we passed the manager's house, on our way home after the movie, we saw Barbara Lakes in her long formal gown, wearing an apron and washing dishes together with her Ladyship, in her gown, drying them.

Barbara Lakes, circa 1969. Courtesy of Peggy Green (niece of Barbara Lakes).

Harold Lakes was an American, a "diamond in the rough" kind of guy. He said, "Now I've got to go home and put a boiled shirt on, just to have my dinner." I don't recall how long they were at the mine—probably less than a week, but it was quite an event. Mr. Banks was charming, a very short man, but very English. His wife was nice too, but an authoritarian. I don't think there was any question of who ruled the roost in that household. Can you imagine, coming up to a mining camp and expecting everything to be as it was in Government House in Victoria?

Barbara Lakes later had a big tea party for the wives and everyone had to get dressed up, with hats and gloves etc., even though the spring melt water was running and the roads were muddy. I remember getting ready for the tea with my son in my lap.

We were all a bit nervous, but she was very nice. Then she started telling a bunch of off-colour jokes. I couldn't believe it. I knew Barbara Lakes before I lived at the mine, for I did a lot of singing in Nelson, and Barbara was in the dramatic society. She was a real character. My grandfather was the landscape gardener who did her home.

I was in the office the day Harold Lakes had his stroke and Gerry Gordon had to take over. Gerry was a great guy and had been Harold's assistant. He was a riot. He had been a major in the army and he had his kids all trained. We were at a convention in Spokane and the Gordons' family came along. One morning in the Davenport hotel he had the children all line up in the lobby. "Number one, you do this. Number two, you do that. He was very nice about it, but they were all lined up in regimental order. I think he had four children. I worked for him until my first boy came along, after which I was too busy at home.

The following are a few other memories I have from my Canex days:

I remember Lawrence telling me that he walked into the Ymir bar the day after the July 1 celebrations and found the floor still covered with beer up to an inch deep. Tony Triggs, my husband, and a lot of others had to get hearing aids because they had no ear protection. His wife, Nora Triggs, nursed my mom in the Nelson Hospital. My nephew, Bill Stringer, ran the butcher shop in Salmo for a long, long time. His six kids were all athletes. Phil Graham, a mine accountant, and my husband used to play on the Salmo ball team. Andre's [Orbeliani] mother and my sister-in-law in Nelson became close friends. My sister-in-law was a French teacher who went to Mrs. Orbeliani to improve her French skills. Andre Orbeliani was a real gentleman, not only his manners, but as a person too.

There were two Finnish miners at the mine who did not get along with each other. They were bosom buddies until they had a few drinks, after which they got into terrible fights.

Canex moved us to Arizona for five years and from there into the Vancouver office. After Joe moved to Vancouver, he spent his summers in the Yukon and his winters in the Philippines. He worked all over the world including Papua New Guinea, but I stayed home and looked after our children.

PAUL KOOCHIN

I have known Paul since the fifties, but not well. His sister married my good friend Bob Chisan, a former work mate at the Dodger 44, who was killed at the Craigmont Mine. Paul was at the Emerald from 1949 to 1954 and is remembered by many as a great raise miner. He now spends every summer on his property on the shores of Kootenay Lake, but can't get in and out of his boat anymore. He had been waiting for hip replacements for seven months when I saw him in August of 2007.

I was born in April, 1932 at Shore Acres, about 20 miles from Nelson. My dad worked mostly in sawmills, but he also worked at the Sheep Creek, Gold Belt, Queens, and the Second Relief mines.

During the late forties I worked three summers as a flunky in the Emerald cookhouse for Ace Bailey, who was a good cook but not the cleanest, and he was drunk much of the time.

I next apprenticed with Canex as a lineman on the power line to the mine, but quit that when I learned that miners were earning twice as much plus bonuses. I worked in drifts and raises at Canex as well as operating Dart and Euclid trucks. I started underground in 1950 and worked with Chris Christiansen. Tony Yosky, Haywire Nick, and "Swede" Ekstrom as a helper. Chris was big and strong. He could pick up a 45-gallon drum of diesel fuel and throw it into the back of a pickup truck by himself. However, you wind up paying for things like that just as I am doing now *[Paul has severe arthritis]*. I used to carry two 50-lb. cases of powder up the raise instead of one at a time. I worked at the Emerald until about 1954–5, after

Paul Koochin of Salmo, who spends his summers fishing in Kootenay Lake near Boswell. 2007.

Larry Jacobsen in 1952. Wally and I met at Canex. He worked in the Jersey, and I worked in the Dodger Forty-Four. We became buddies and for a while shared ownership of a 1936 Plymouth sedan. We were neighbours at Remac after we both married in the mid-fifties. Wally's older brother Terry married Georgina's younger sister, who had five sons by the time she was twenty-one. Courtesy of Wally and Georgina Bunka.

which I worked for Custom Rock and Bulldozing in Salmo. It was owned by Bob Rotter, who was very good to work for. I had worked for his father Frank Rotter too as a kid, and my dad worked for him as well.

I was in that Dart truck that Bill Peters drove to Salmo for beer in 1952. I got away with it because I was young and innocent. Just about the whole crew went with the truck to town. A lot of the crew was fired for that caper, but then they called them back. I didn't even get a reprimand because I was only 19 then. The Peters boys, both Bill and Avery, were a lot of fun. Wes Peters was also shrewd and a good poker player.

At the Emerald it was all two-man raises while I was there. Terry and Wally Bunka both worked in the conveyor tunnels, as did Johnny Larsen. Larsen was later killed on the Island, when the mountainside caved in and took him and his airtrac [*a self propelled track mounted drill*] into the ocean. Wally once drilled into explosives. They went off and he is now blind.

During my mining days I drove a lot of raises, especially at Remac, where I worked following the labour strike, until they closed in 1975, in addition to two years at Canex. At Remac I drove one raise 800 feet by myself. R. F. Fry & Associates, the contractor using an Alimak raise machine, beat me by only a couple of days. It got to be a heck of a long climb towards the end. I also worked for Bob Golac a total of about 17

Wally Bunka in 2007. In addition to mining, both he and his brother Terry have been stock brokers. Wally has probably made and lost a couple of fortunes. Because of losing his eyesight, he now finds it difficult, but not impossible, to track his stocks. Once mining gets into ones blood, it often stays.

years—much longer than I had worked at Canex. My last stint underground was at Kemano in 1990–2001.

I remember Ken Henderson, who was an accountant at the mine. His daughter Ann Henderson became the mayor of Salmo. He also has a son who lives in Salmo. Stan Hunter was at the Emerald when I started there. He was a good man. I also recall that Andre Orbeliani wrote a constant stream of letters to the editor of the Nelson News.

We had a good ball team at the mine during the fifties. We used to go down across the border to play the guys down there. One Saturday we were going across the line to play. When the US border guy at Nelway asked if we were all Canadian, the bus driver Jim Unruh said, "No. My wife's a DP from Montreal." We had to turn around and go back to Salmo to get our birth certificates. Those guys had no sense of humour then, and they are probably worse now.

My house in Salmo has a full-size basement, and I wish now it didn't. The washer, deep freeze, and wood heater are down there, so I have fourteen steps to descend and climb each time I go down there, and I normally go through about seven cords of wood in a winter. When we moved into that house 48 years ago, oil was $0.18 per gallon. I said at the time, "I will never even look at a block of wood as long as I live." Now it costs about $800 each time I fill the tank.

Georgina and Wally Bunka, circa 1952. Wally, like me, did not work for Canex very long. He did go back briefly in 1953 to work in the conveyor system tunnels. He lost his vision many years later when he drilled into unexploded dynamite in a bootleg. Courtesy of Wally and Georgina Bunka.

MEL OLSON

I interviewed Mel in the house he shares with his partner Audrey J. Leggitt in Maple Ridge (his wife is deceased). Mel provided a wealth of information, much of which is reprinted in the appendices.

Mel was born and raised in Saskatchewan but came out to BC just after the Second World War to meet his brother. Finding a job may have been relatively easy, but finding a job that suited a young man wanting to enjoy life was not. Mel moved from Lytton—"the hottest place in summer and the coldest in winter," where he was a lonely assistant agent with the CPR—to Penticton, dispatching boats, then serving as the assistant purser on the SS Minto, not having enough time off. His next job was at the Cominco lead plant in Trail, which was not good for his health. After that he felt exploited doing contract loading of fertilizer bags together with "smart old buggers."

Mel says, "Getting into the workforce when I did ensured me getting a healthy work ethic. It was either that, or starve."

He did indeed develop a healthy work ethic. In addition to his regular

Mel Olson, CGA, with his partner Audrey Leggitt, 2007. Mel was a long-time Canex warehouseman, accountant, chief accountant, Scoutmaster, etc.

work, he became prominent in the accounting associations and has held important elected offices there as well as in chambers of commerce, service clubs, and other volunteer organizations. He was a member of the first Canex team to win the West Kootenay Mine Rescue Competition Shield. At Canex he was a Scoutmaster and he was also in charge of producing a small company newspaper.

His children have learned well too. His son Gary is a staff sergeant with the Toronto police force and his son Mark is VP of Canadian Helicopters. He also has a daughter who lives in Coquitlam.

In 1950, at age 22, Mel was hired by Canex. Here is what he told me about his time at the Emerald:

I immediately got a place in the bunkhouse. I will never forget the old cook, nor some of the girls who worked there.

As far as accounting is concerned, Colonel Perry was the guy that kept the early mine controllership post together. He'd call you in and ask you to do something and he didn't care how you did it as long as you did with a reasonable job that he could present to management or operations.

He called me into his office one day and said, "Mel, you can buy all the shares of Placer you want. You're a young man and it's a good time to buy them. We are giving you a chance to buy them at one dollar per share. We'll take them off your paycheque, one share off each paycheque until they are paid off. How many shares would you like?" I said, "Colonel, I came here and my wages were this amount. The cost of the bunkhouse was that amount, likewise, the kitchen. When I'm finished with my car payments and you give me a $20 raise and think

Main camp: three dwellings, main office, kitchen, and two bunkhouses. Early 1950s. Courtesy of Marie Adie.

you have done me fine, I want you to know that the price increase of my board is higher than the pay increase and I've got to pay tax on the $20 on top of it all. Do you think I'm going to invest in the company? I'll work for them because there's no better place to work right now, but I'm sure as hell not going to invest in Placer." He was absolutely dumbfounded. In retrospect, if I had invested in the company then, I would now be rich.

Then I bought a car from Columbia Motors. They needed to do some work on it before letting me take it, so the garage gave me a loaner to use to take my scheduled driver's licence road test.

Main office up close. Courtesy of Mel Olson.

The examiner at that time made everyone taking a driving test in Trail drive up the fire-hall hill, stop on it, and then start again. The emergency brake on the borrowed car turned out to be defective. When I stopped on the hill, it would not hold the car, so as soon as I took my foot off the brake pedal, the car would roll back before I could get my foot on to the gas pedal. I failed my drivers test because I had not checked out the car before doing the test with it. I wonder what the manager, Legg, would have done if he had known that I was driving him around the mine with no driver's licence. I had done lots of driving out in Saskatchewan and had lots of experience with vehicles.

I began at Canex in 1950 in the warehouse. Then I quit and went to Uranium City, where I became the chief warehouseman. When I returned for my second stint as warehouseman at the Emerald, I got married, and that may have had something to do with me settling down. I also got a company house and good money.

I worked in the warehouse, but also studied accounting and eventually got my CGA. I then left the warehouse and became a "mine accountant." Years later when the company was installing computer systems, I was sent back to the warehouse because it was the most problematic as far as computerizing was concerned.

From there I moved into Cost Accounting and became involved in running a manual system parallel to the computer system to ensure the latter was working as designed. I also went to the warehouse and got the inventory onto the computer system and running properly.

I went to Craigmont for a while, but the "Salmo Mafia," as we were called, was deeply resented there. They gave me First Aid people off the prison farm to train in warehousing, but that was a hopeless task. It was an impossible challenge to teach green staff how to keep records and to store things where they could be found again. I tried to convince management that they needed a warehouse before general construction began. Once we got behind it was impossible to catch up, especially with people who could not even run an adding machine. I did not like it there, and Gerry Gordon made sure I came back to Salmo.

Six months after returning to the Emerald, I was the Chief accountant there. I stayed there until I transferred into head office a year or so before the mine shutdown. I had to get out of there because my kids were of an age where they needed to further their education.

Dick Ash was working for us, but his wife was sitting at home and having a difficult time of it and was sick a lot. I needed a girl on the switchboard, but it was not a great place to work because of the static from the frequent lightning storms. We were right in thunderstorm alley and lightning was often a problem for us. Once my wife was hanging out clothes, when lightning struck the clothesline and arced right up to the house. She was scared spitless to go near the house. At the telephone there was little or no protection, so there would sometimes be some flashing during a storm.

I said to Dick Ash, "I need someone on the switchboard, and it might be good for your wife to get out of the house, so why not have her try it out. Maybe an interest outside of the house will make her happy." Well, she came to work, and she not only looked after the switchboard but helped edit our newspaper and did a lot of other things as well.

Earl McLean's wife, Corrina, was another bored wife sitting at home. She had had breast cancer, but she came to work for us and turned out to be great. In addition to her other work she printed most of the newspapers that we published.

Dick Ash, Dale Owen and an unknown person. Courtesy of Sultan Minerals Inc.

Gerry Gordon was the one I liked the most. He came up to me one day and asked me about two little sheds behind the bunkhouse where they stored the explosives. When the mine was small they had blasting caps and fuse in one of them that was about the size of an outdoor privy. There was no room in it to work, so the fuses were being capped in the dry where the men were changing clothes. The other building not far from it was beyond belief—stacked to the ceiling with dynamite. It was an incredible situation. The last shipment that came in was hauled underground and stored in a dead-end tunnel because there was no room in the powder house, which in any case was far too close to the camp buildings. Within a few days despite all the other construction work going on with building the mill and other construction work, Gordon had a crew build a new powder house and cap house. He had enough savvy to realize that they could not carry on the way they had been doing. He told me that he was earning only $325 per month when he first came to the mine as manager.

Many of the young engineers that came to Canex spent a month or two helping in the accounting office. I often wonder how much this later helped them in their careers, having a sound grasp of the economics of an operation, for some of them became very successful.

Ken Henderson first worked in the Lead-Zinc office. When we started working on the Emerald Tungsten and getting the mill going, he was moved to the new Tungsten office and became the chief accountant, and I was the chief warehouseman during construction of that mill. I respected Ken. He was a decent person who recently left Salmo to live in a seniors' lodge at Castlegar.

The Panabodes were nice houses, but after a while when the logs dried out, they shrank. There would sometimes be gaps between the logs. You could sit on the throne in the bathroom and if the wind was blowing the snow, it would come right between the logs and dust you. If the wind was really strong, the whole floor would get covered. There were no attics in the Panabodes, so the company sent men around to install liners in the rooms to conserve the heat.

Salmo, a village with a population of under a thousand people and few single women, was ill prepared to cope with as many as two thousand miners and loggers looking for female companionship on their days off. There were stories about "magazine sales girls" (prostitutes?) arriving on Friday evening, staying overnight in the bunkhouses and having to make quick exits before the police arrived on Sunday morning.

There were whispers about sex on the potato sacks in the kitchen store room.

Panabode house. Courtesy of Sultan Minerals Inc.

And a milk delivery man was in shock after getting verbal instructions from the nude housewife at the door.

Some miners would leave the mine with their paycheques on Friday evening and return penniless on Monday morning. Two miners, who were quitting, were waiting at the pay window for their cheques, while the staff was all bitching about how much money these guys were making. When the manager asked what their problem was, one of

Mine Rescue certificate. Courtesy of Mel Olson.

them said, "I gotta get out of here. I gotta get out of here now, 'cause I'm drinking myself to death with all the money I'm making."

The auditors got excited when they discovered a horse on the Canex payroll. It turned out that a small contractor supplying timber for the mine wanted $5.00 per day for his horse and its feed. To get around the red tape and the inconvenience of billing for it, the contractor asked for the horse to be paid as a person with the hours adjusted to cover the expense.

A couple of miners seemed to trust the mine more than the bank. Staff had to beg them to cash their pay-cheques so they could balance the books.

Curling rink. Courtesy of Mel Olson.

Some, but not all of the small pits on the hillsides were dug by Chinese prospectors. Others scoured the area as early as the 1880s.

The Emerald was a lovely place to raise children where they lived and played in natural surroundings. They had a local grade school, good fishing, Boy Scouts and Girl Guides, a swimming pool and an outdoor skating rink as well as the mountain side to play on. The adults had a curling rink, commissary and a community hall for dances and badminton as well as many other activities.

JOE MONTGOMERY

Joe turned 80 a few days after I met him, but he appears to be far too young to retire. He not only looks much younger than his years, but he is still very much involved and was going to Colombia a few days later to consult for a mining company there.

I worked at Canex for three winters as an assayer from 1950 to 1952 and spent my summers prospecting with my friend Lehto.

At the mine I worked in the tungsten lab on top and in the lead-zinc lab down by the highway. I believe Lou Stard was the assistant superintendent at the lead-zinc mill and Grimwood was the lead-zinc mill superintendent.

We had an accountant who had come from a silver mine up north. He was repairing a kayak that he intended to use in the annual race on the Salmo River. He had damaged it when he first put it in the water, so I was helping him rebuild it. He told me that I could enter the race with him, for it was a two-person kayak. I helped him for quite a few nights putting this thing back together. He then decided at the last minute that he was going to get another guy, an older friend of his from town to ride with him. So I said, "Okay, that's fine." Anyway, on the day of the race they put the boat in the water, got about a hundred yards downstream, and hit a rock. That was the end of the race for them. It was the end of that kayak as well.

Joe Montgomery, B.Sc., M.Sc., Ph.D. (Geology), still works at age 80. 2007.

I used to go swimming in the river in the swimming holes down below the mill. There were tailing ponds all around, but the water was okay there. It had nice pools and I used to take my dog down there.

I wish I could remember that accountant's name, for we used to go into the Salmo pub together once or twice every month. The accountant would first stop at this little coffee shop where he would buy a can of Copenhagen snuff. In the pub he would open up the can and put a pinch of snuff in his mouth and then start drinking beer. By the end of the night, there would be no more snuff left in the can. He would have swallowed it all with his beer. That really impressed me.

Ken Lehto and I worked in the labs during the winter and then took off in the spring to go prospecting. We would

Joe with his partner, Frances, in 2007.

then come back in the fall and try to get a job again. One winter we worked driving underground trucks while we were waiting to get into the labs again. Once when I came back, there wasn't that much work in the lab for a while, so I was a diamond driller's helper underground. The noise was just unreal. It had a penetrating high-pitched whine. After Salmo I worked as a helper on rock drills as well including a time with a Copco jackleg. I had also helped on a leyner drill for which we first set up a bar and arm on which to mount it. It was heavy!

I was born in Vancouver, but we moved to Moose Jaw when I was six months old and lived there until age fourteen. I then moved back to Vancouver with my family, finished high school, and enrolled at the Vancouver Technical school on Broadway near Nanaimo Street. It was strictly a boys' school, and I never met a girl until I was about 21. Ken Lehto and I went to high school together and became good buddies. When we graduated, we both got jobs for a couple of years with the BC Research Council as lab technicians. We decided that we didn't want to spend our lives as lab techs for low wages, so we did some research. We had little money, so we looked into the main industries of BC, which were forestry, fishing and mining. We decided that we couldn't afford any lumber or buy a fishing boat, so we went prospecting. We both took a night course at the university that Harry Warren used to put on. He was a very well known professor out there. So we went prospecting in the summer and had the lab jobs during the winter. We did find some stuff, but nothing very exciting. At that point I decided to become a geologist and he a pharmacist. So we both went to UBC at the same time. Ken graduated and opened a drug store in Kaslo, where he stayed for a couple of decades before retiring and returning to Vancouver. I graduated from UBC in 1959 with a bachelor's degree, in 1960 with my masters, and in 1967 with a PhD.

I had worked in a little gold mine in the Tulameen after my job at the Research Council. I did everything there. I ran a little mill, helped the miners, drove a dozer, etc. The mine collapsed financially. It was December and they asked me to stay there until the snow came. So I said I would stay there and then ski out. I stayed for about a month as the snow got deeper and deeper. Next they said they couldn't pay me, but told me to take whatever I wanted. I took mostly books and things like that. There was this Chinese guy who worked up at the Silver Hill mine way up the road from us. Before the snow came, he came by with his truck and I put a whole bunch of stuff on it and took it out and sold it. I got paid okay! That's when I went to work at Canex. When I worked at the tungsten lab, it tended to keep me isolated from the others. I lived in the miners' bunkhouse. I like miners. They've all got that look, "I'm having a good time. Piss on you."

My job was to analyze the cons *[concentrates]* for about 10 trace elements. They had to meet very strict technical requirements. They had a lot of problems with their molybdenum. They couldn't get the impurities out, so they kept running it back through the mill and do more and more things to it until they got a really clean concentrate. I would then do my analysis again for the 10 trace elements and give it the okay. Because of my work there I later got a job with G. H. Eldridge in their Vancouver lab.

They did a lot of different things in that mill. They started out with big shaker tables, where they run water across using a gravity separation process, and couldn't get a clean concentrate. By this time the tungsten prices had started going down, but they had a long-term contract with the US government for $60 per unit (20 lbs). Canex was able to keep selling their product long after the other tungsten producers had shut down. The US had to keep buying their concentrates, but they got very sticky about the specifications.

I used to sit and chat with the night cook at the lead-zinc mill. We'd sit and sip coffee while he was baking pies. He was an exuberant chap whose favourite trick was to keep a bottle of Crème de Menthe in his pocket when we went to the pub. He would take your glass of beer and drop in a bit of the liqueur under the table from the bottle in his jacket pocket. It tasted terrible, but after a few you no longer cared.

Banks came to visit the tungsten lab. He was sort of like royalty. You could tell he had picked up a little knowledge about tungsten labs and assaying in general, because he asked questions that a normal person wouldn't ask. I could hardly understand him for he had such an English accent. I had to really listen.

Years later after becoming a geologist, I started a prospecting syndicate together with Angus McDonald. So we went looking for money to finance it. Joe Adie worked for Canex then and was in charge of looking at submissions brought to the company. He was a great guy. We got Canex, Noranda, and Homestake to back us. We did that for four years. I got to know a few of the Canex people during that time, and Joe Adie was just a great guy to do business with. He was a real entrepreneurial type who was not afraid of taking a chance. He was the guy who got Canex the large Endako moly mine after 20 other major companies had turned it down. Joe Adie said, "Well let's stick a hole in there anyway," and their first drill hole hit the ore. The earlier holes had all missed it. That mine has turned out to be a real money maker.

While I did not work for Canex that long, I had a lot to do with them over the years. They drilled several prospects that we brought to them, including the first porphyry copper deposit found in the Yukon near Carmacks. It was large enough, but too low grade to be commercially viable. Fred Gouder was the guy who owned it, and he is some story, that guy. He came from Germany as a teenager all by himself and went prospecting.

I used to do hand-steeling when I was prospecting. Ken Lehto and I worked way up on the mountain on Blackstone Creek. It was hot up there in the summer and we'd haul water up there, not to drink, but for hand-steeling to make mud. We'd pour water down the hole. We sank a little shaft by hand. It was about five feet by five feet and perhaps six feet deep in a really rich silver-lead-zinc vein. The vein was four feet wide, but it tapered down to nothing in a few feet. It had all been eroded away, and that's how we found it because the mountain side was littered with float *[rock containing ore]*. There was solid but oxidized galena *[lead-carrying ore]* all over the hillside. We sank that shaft by double-jacking. I think we used about a 12-pound hammer. We took turns swinging the hammer and we were extra careful, you know, because you got to hold the steel when the other guy swung the hammer.

EARL MCLEAN

Earl McLean was 87 when I met him. He and his present wife, whom he met on a dance floor, go ballroom dancing regularly. Earl McLean was kind enough to lend me a copy of Placer's "25 year club" edition of Placer News *from 1976, which had photos of many of the people who had been at the Emerald.*

The way I found Earl McLean was sheer coincidence. I was talking to the Assistant Librarian at the Princeton Library and mentioned I was writing a history of the Emerald. A woman sitting at a nearby table piped up, "I was born at the Emerald." She turned out to be Earl McLean's daughter, Sheila Dixon, whose story also appears here.

I was born in Kamloops, grew up in the Shuswap area and came to the Emerald in February 1950 as an electrician. We paid $17.00 per month rent for our house, including the utilities, and got the heating oil supplied free, too. Ours was one of the first families to move up to the townsite. They hadn't even backfilled the utility trenches yet; it was November, and it had started to snow.

The Canex order for Panabode houses was the first big order that the Panabode company got. Those houses were beautiful but draughty and were equipped with oil furnaces that could be cantankerous. You had to watch them, for they would sometimes flare up and blow soot into the room.

In most mines staff does not fraternize with hourly paid people, but at the Emerald we tried to get away from that and tried to minimize the difference between staff and miners.

Earl McLean was the electrical foreman at Canex for many years before moving to head office. At age 87 he was still dancing weekly—not weakly. 2006.

The company provided the community hall—I wonder if it was an excuse to get us really smashed sometimes. The bonspiels were of course big drunks. The swimming pool we built is still there.

One important reason the Emerald survived was because we were able to switch back and forth between the tungsten and the lead-zinc. When the lead-zinc prices were poor, we mined tungsten, and vice versa. The mill at the bottom was originally tungsten, when Canex took it over, but they switched it to lead-zinc two years later. Then when they started mining tungsten again, they built a new tungsten mill up on top. We had a tram line to haul the ore from the mine down to the mill below. Cory Reyden looked after it while it was operating. It was quite a thing in its day, but it gave him many headaches when the buckets sometimes collided with the towers during strong winds.

My electrical crew varied in size, usually seven to ten men. We not only had the mine, mill, crusher, and conveyors to maintain, but we looked after the telephones as well. We even rewound some electric motors, although with big ones, we had a crew come up from Vancouver to do it.

Canex was the first mine in Canada to go with trackless haulage. There were lots of larger mines back east, but I guess they were still using tracked equipment. I guess even Kimberley was all conventional mining. I think Canex was the first small mine to hire an electrical engineer, and a mechanical guy. Previously in the smaller mines, the mine superintendent was God Almighty. He did everything. Canex did away with that when they hired engineers. Before that the mine superintendent wasn't necessarily an engineer either.

Dick Grimm used to truck the ore from up above down to the mill after they had changed it to lead-zinc. After we set up the conveyor system, we no longer needed him, but he continued to haul ore at the mine for quite a while, even if it wasn't to the mill.

The company had built a two-mile pipeline over to Lost Creek to get water for the mine. I went out there one day, and here was this whitetail buck—a guy had shot it, knocked it down, but only wounded it. When it got up and started to run, I shot and killed it. The guy came over and argued that it was his deer since he had shot it first. We finally tossed for it, and I won the toss. There were often deer right in our back yards. We had bears up there too—a lot of them.

We used to go hunting down in the Flathead and had to go in through the States because there was no road in from our side. Once when Dick Grimm and a bunch of us were down there, we got one too many elk. Coming back, we had to cross the border into the States and then back into BC, so we cut the one elk up into fairly small pieces and put it up underneath our camping gear and got it home. We always went there late in the season—in November, after we had at least a foot of snow.

When Nick Smortchevsky moved to Vancouver, I came down and became his assistant. I ordered all the electrical supplies because we designed all the mines from Vancouver and shipped the materials to the sites. We marshalled everything for every mine we built. Mechanical had to do the same. We had quite a few mechanical guys here.

We didn't have a ski hill at the Emerald, so when we got tired of driving the kids up and down the hill to Salmo to ski, we started skiing ourselves. A bunch of us from the mines built the Salmo ski hill after we first cleared the land.

We had a month-long labour strike in 1959 while Gerry Gordon was still there. He was a tough character. After we settled the strike, he began charging the men $0.10 a day for their bus rides.

We had a lot of good men work at the mine over the years, and Ed Lawrence was our last manager. My wife was his secretary at the end. Ed said that the hardest job he had was making money. I remember Grutchfield well too. I retired in 1985 but went back consulting for the company. I worked 23 years at the Salmo operation and almost 20 years in the Vancouver office afterwards.

SHEILA DIXON, NÉE MCLEAN

I interviewed Sheila at her home in Princeton, BC, where she now lives. Her father, Earl McLean had been kind enough to give me her phone number.

I was born at Lethbridge, but came to Ymir in the spring of 1950 when I was five months old. We stayed there for a few months until the townsite was built. We moved up to the mine that fall and were one of the first families there. My father was an electrician but later became the electrical foreman.

I went to school at the mine until I finished grade six. The school had two rooms, grades one to three in one room and four to six in the other. From grade seven on, we were all bussed to the Salmo high school.

I had a wonderful teacher, Miss Owen from Nelson, the first year, but at the end of the term she got married and left. I liked our principal, Mr. Kinakin, who had a daughter who was in my class during grades two and three. After Miss Owen left, we had a teacher named Carol. She was young, inexperienced, impatient, and prone to losing her temper. One day she threw a blackboard brush and accidentally hit the principal's daughter on the head. I don't recall any other incidents. She married a miner and stayed up there, so we passed her house every day on the way to school.

Sheila Dixon, 2007.

In those days there were many young families at the mine. When the parents got together for a party, my mother played the piano, and the Salmo music teacher, Don McDonald, played the saxophone. Mr. Rydall was also a teacher at the mine who played the clarinet. They didn't play rock 'n roll yet—it was mostly wartime or late '40s music. The parents often had events at the community hall, and they would get one babysitter to mind a bunch of us. I still have a clear mental image of my mom getting all dressed up to go out. For years they did a lot of square dancing. It was lovely having that community hall, which they used for parties for the kids every Christmas and Halloween. The mine sponsored everything.

Years later, a bunch of us were in a big restaurant in Dawson Creek, sitting next to a table of doctors who were listening to our conversation and yelling comments at us. It turned out that one doctor was Carl Wilson's son *[from Salmo]*, who had academic difficulties in high school until he decided to become a doctor. He then breezed through university with no problem at all. It's remarkable what motivation can do for a person. He practised in Texas for a few years before moving to Dawson Creek.

Growing up at the mine as kids was wonderful. In the summer we would just bring a little bag with a sandwich and take off on our bikes, and we didn't have to be home until dark. I remember Dale Clayton and Brian Anderson. The three of us would ride all over the place on our bikes, to Indian rock, or to a little creek down on the south road. In the winter we tobogganed and skated. I started taking swimming lessons when I was three years old. We had a swim teacher/lifeguard every summer. She usually came from Trail, and the company provided a house for her. We had to spend two to three years in each swim category, for the age limits applied to us all. We got to be like fish—we swam so much. We used to compete when we got a bit older and had meets between the schools in Salmo, Trail, Nelson, as well as other places in the Kootenays.

When I was nine or ten, we moved into a duplex at the end of the road, from which we had such a great view down over the Lost Creek valley. It continually amazed me that I was actually looking over an international boundary and into another country. The population at the mine had declined, so we got both sides of the duplex, and Dad used part of it for a workshop. However, it was scary coming home at night because there were so many bears in the area. There was a little cinnamon-coloured bear that put its paws on my window sill and looked in at me sometimes. It was the cutest little bear and I adopted it and called it my bear. You can imagine how shocked and hurt I was, when, coming home one evening, I saw it hanging from a tripod next door at Al Nord's place. Years later when my son was graduating from high school in South Slocan, during a break in the ceremony we all went to the neighbourhood pub. This guy came over to our table, nodded to Dad and said to me, "I'm Al Nord, your old neighbour. I've still got your cinnamon bearskin hanging on my wall in Krestova."

Pelt from Sheila's bear on Al Nord's wall, 2007.

I remember one evening heading down the path to meet twelve-year-old Carol,[8] who was coming up the path when I hear her scream. I looked up and saw this white bear walking on its hind legs. It looked huge. Carol ran back down the path when another white bear came towards us from the direction of my house. Wildlife officers came out and trapped them, but I never found out what they did with them.

Our life at the mine was somewhat insulated from the outside world. That changed when I began school in Salmo. I met a lot of kids down

---

8 She is now Carol McLean. Her story appears elsewhere in the book.

there, but my dad did not let me hang out with them and insisted I come home on the bus every day. Somehow I managed to miss the bus occasionally, but it was never long before my dad came looking for me. I did make friends with a few girls who would sometimes visit and stay overnight. I would sometimes be allowed to stay a night with them as well.

Dad sent me to Penticton, where I took my grade 12 and worked on Saturdays. I got my room and board at a private home. I can understand now how he felt. Before he sent me away, I had been in three serious car accidents.

Dad was always health-conscious and would drink socially at a dance or at a party. He might occasionally even have a drink after work, but never in excess. He was stable and emotionally steady. Without him I would never have made it growing up. He was a wonderful father.

After Dad retired, he went back consulting with Placer Dome. He gradually worked fewer days, until he was down to two days a week and then finally retired for good in his mid-seventies.

When the mine closed, Gilbert and Carol Wilson were able to get one of the Panabode houses from the mine and move it into Salmo. It has these curved doorways and is really beautiful both inside and out. I saw it during the Canex reunion in the mid-nineties.

Dick Grimm had a trucking business at the mine and was my dad's hunting buddy for many years. They went to the "Flathead" area in the east Kootenays every year and stayed until they had an elk and a deer each. When they came home my dad would have a big elk rack on the car and we would all be out hooting and hollering to celebrate. Carl and Ethel Wilson had a service station in Salmo, where Carl and Dad hand loaded their own rifle cartridges.

Sheila McLean as a teenager. Circa 1967. Courtesy of John Bishop.

One of our favourite picnic sites was across the border at Sullivan Lake. Canex also sponsored the annual company picnic there, where we played ball, ran races, swam, and generally had a great time. I drove down for a visit in 1990 and was relieved to find few changes since my childhood. The reason, I guess, is because the lake is within the Colville National Forest.

While I lived at Slocan Park, I was an "activity" coordinator at the Willowhaven in Nelson, as well as at Raspberry Lodge and Castleview Care centres in Castlegar. There I got to know my neighbours well because I often looked after their parents.

JEWEL OF THE KOOTENAYS / 101

People would come to visit a relative or friend, and it would frequently be someone I had known as a child. I found that very nice.

Linda Thomas was in school with me and her father Rae was with the Paraplegic Association. He had once been invited to a meeting at Nelson to talk about what he thought the city needed, but he could not attend that meeting because there were no wheelchair ramps. He delivered a statement on the front steps of city hall instead. In 1998, after my oldest son fractured his neck and became a quadriplegic, he moved back to Nelson because that was where his friends all were, and because of Rae Thomas the city is now wheelchair-friendly. So a belated thank you to Rae.

I feel so lucky to have grown up surrounded by nature as we were at the Emerald. We were not restricted the way children are today. It has given me a love of nature and the outdoors that I would surely not have if I had grown up in a city. I am used to having five or ten acres, a huge garden and few neighbours. I now live in a subdivision for the first time in my life, but life is good. The mine was wonderful to grow up—even though we were in the bush.

CLIFF RENNIE

I interviewed Cliff in late summer of 2007 at his Nanaimo home, where he now lives alone. He was at Canex from 1951 to 1960, and from then on in the Vancouver office whenever he was not in New Guinea or the Philippines. He is another one who does not know how to retire. He is still the chairman of a small mining company despite reaching 80 years of age in 2008.

2007 photo of Cliff Rennie. At age 80 he is still involved in mining.

Cliff grew up on Vancouver Island north of Courtenay on what he calls "a farm in the bush" and at first had "no interest in rocks," but that changed once he went to university. Here is Cliff's lively story, and while not everything is directly related to the Emerald, Cliff's experiences in the world of mining seem relevant enough to include them here:

I went to UBC together with Doug Little, Tony Triggs, and some war veterans and graduated in 1950, but because I took a couple of supplemental courses did not get my degree until 1953.

At UBC I was rooming with a guy who was

taking a course through the Department of Veterans Affairs (DVA). He had worked at Wells, so when I was looking for a summer job, he said, "Why not come up to Wells and work in the mine? They're looking for people." So I did.

I went to the Cariboo in 1948 and worked at the Island Mountain mine. Working in the gold mine got me interested in rocks, minerals, and geology. Prior to that I had always liked building things, so I had leaned towards civil engineering. I had also taken a correspondence course in electricity during my high-school years, but found that I wasn't that hot in mathematics, so electrical engineering was out, but geology did not require great math. Meanwhile there was a geologist named Tony Barker one year ahead of me who was working for American Metals. They had a concession up at Great Slave Lake, so I ended up going there to help map it, but there were no outcrops at all, just tree-covered flats. We both wound up as diamond drill helpers. I worked with the diamond drillers and acquired diamond drillers' attitudes. The driller I worked with wanted only to drill five feet more than the cross shift had done, after which he'd roll up in a tarp and have a nap. This was when we were on night shift, but I did not have tarp to roll up in. Anyway, I got to know what was going on in that end of the business.

Doug and Jean Little's wedding. Mid-1950s. Courtesy of Marie Adie.

The following summer (my last as a student), I got off on oil and gas surveying in the Monkman pass area. I was chasing and mapping gas seeps. I learned that the grizzly bears liked the gas for some reason. They would dig a big hole at any seep they found, which would fill up with water and they would bathe in it. All we had to do to find these seeps was to follow the grizzly bear trails.

After I graduated, Charley Ney, who was chief engineer, chief geologist, and mine superintendent at the Kicking Horse mine at Field, BC, hired me as his assistant. I had been working there for a year and could see the end of their ore reserves. I now had a wife and a baby and lived in an apartment in an old teahouse in the northeast corner of Field, where the wind came from. I was getting $175 a month and couldn't afford the oil to heat the apartment. At this time the price of lead and zinc was up, so half the miners quit and went to Salmo. Canex was advertising for geologists, so I wrote them a letter, was offered a job, and I was soon gone too. While I was still at Field, a dozen miners came in on a Monday morning and quit. The manager confided to me, "When all those guys get up and quit, it makes me feel like quitting too."

I came up to the Emerald on October 20, 1951, and stayed in the

The Rennie clan: Two grandparents, six sons, four wives, twelve grandchildren, Christmas 1996. Courtesy of Cliff Rennie.

Mrs. Rennie holding the first two sons. Courtesy of Cliff Rennie.

Cliff holding the first two sons. Courtesy of Cliff Rennie.

mine staff house at the main camp for a little while. In what was called the shack camp, were three log cabins and some frame buildings. I learned that one of the log cabins was vacant, so I asked the chief geologist, Clive Ball, about moving into it and he said, "Okay."

I moved into the cabin in the end of November. I ordered an electric stove, pulled it in on skis, and hooked it up by myself. When I turned on the oven and all the elements, it gave me 3,900 watts of heat. I lived in that cabin until they finished the Panabodes, and I was able to move into one of them over in the Jersey townsite. It was nice. The house had an oil furnace (just like a Coleman Heater) suspended below the floor and it threw enough heat to keep the place warm.

My wife grew up at Courtenay and I met her at high school dances there. We got married in 1950 and ended up with six sons: Bruce, a Systems analyst with Weyerhaeuser; Matthew, a helicopter engineer with Vancouver Island Helicopters on heavy hauls; Michael, a diamond driller and gentleman farmer; Gordon, an electrician near Sydney, Australia (we had lived in Australia for five years, so he went back and married his high-school sweetheart); Jack, a journeyman mechanic with the Highland Copper mine; and Howard, who was into diamond drilling and was killed last year on October 17 in a helicopter crash near Alice Arm, BC. My wife died two years ago from a cerebral hemorrhage.

It was around New Year's Eve in 1952 and we were at a big celebration down at the tungsten, when my wife suddenly said,

"We got to go to Nelson." She had decided to give birth to our second son. So we were heading out in our little 1940 Austin car, when Corey Reyden, coming home from the party, came running over and said, "You can't go to Nelson sober!" He ran into his house across the street from us and came out with half a tumbler of rye and said, "Here, this will keep you warm."

We trundled off to Nelson and got there at four or five in the morning only to find the door of the hospital locked. We banged on the door until a nurse opened it. She took one look at my wife and yelled, "Here's another one!" We didn't even know it then, but Bob Weber, who lived next door to us, had their baby the same day, except theirs was the second whereas ours was the first.

We had two babies born in that hospital before my wife said, "Enough of that. That's a terrible place with the cockroaches running around all over the place." So she went to Trail for the next one. It was a good thing too, because the baby's hip socket had not developed properly and they discovered it right away. He wound up in his first cast when he was three days old. With the care he received there the problem was corrected by the time he was five months old.

Clive Ball was the son of a prominent geologist in Australia. He had flown the "Hump" during the war, freighting supplies from India to China in unarmed freight planes, and won the Distinguished Flying Cross (DFC) for his work. He was a very nice guy, but not forceful. He was forever apologizing to everyone and everything. In his house he once tripped over a footstool, and then apologized to it.

There were eight geologists at the mine, including Clive Ball. They got me going on surface exploration on things including the Lincoln Clubine gold prospect claims down towards Lost Creek. Clubine was quite a character who had owned the claims jointly with Cominco. When Cominco refused to buy him out, he bought them out instead. As far as I know, nothing has ever been done with this property.

In the fall of 1951 when the Copper Mountain mine near Princeton was closing down, Canex

Mrs. Rennie tossing kids from the step into the deep snow. Courtesy of Cliff Rennie.

Cliff Rennie digging kids out of the snow. Courtesy of Cliff Rennie.

Jack Anderberg, foreman at the Dodger while I worked there in 1952. Courtesy of Marie Adie.

brought over its general superintendent as mine manager. He in turn brought over half of his crew, including the chief engineer and other staff. His brother-in-law, who had been the mine foreman at Copper Mountain, became the mine superintendent.

Jack Anderberg got fired one day when he jumped into the truck, got his men in too, and drove off leaving the mine superintendent walking. The Copper Mountain crew had no control over anything, and it was during their reign that the Dodger 42 crew drove a Dart truck into Salmo for beer. The crew was also using the Eimco overhead loaders to play "catch"—throwing large rocks back and forth between them. One day a loader went off the road and into the dump at the Dodger. Instead of trying to recover it, they were burying it under waste rock. It was not long before Simpson, the managing director, sent the whole Copper Mountain gang packing and promoted Gerry Gordon to mine manager. Gordon and Simpson had worked together during earlier times.

I used to hike from the dry up to the Emerald mine, then up to the Dodger 44, and back down to the Dodger 42 *[2 to 2½ miles]*. One time I got a ride with Hub Maxwell in his panel van. He was wheeling down the road in a big hurry, went around a corner and down over the bank. I was sitting on a bench in the back of the van, whose back doors were open. As the van went over, I hopped out into the snow bank. The van rolled down the embankment with Hub bouncing around inside like a rat in a cage. He didn't hurt himself, but boy was he cross when he crawled out of there.

Joe Adie was at the Emerald in 1948. He is long gone, but his widow Marie lives in North Vancouver. Joe did everything at the mine. That's why he got the name Joe, because he was the Joe-boy for everything. It wasn't his name but that is what everybody called him.

In the beginning they intended to ship 5,000 tons of lead-zinc ore from the open pit to the smelter at Trail, because that was all the lead-zinc they could see. However, they wound up mining 50,000 tons of it in that pit. They made enough money from that to diamond-drill further to the north and found the "A" zone, and from then on they were in the lead-zinc business. Joe laid out the underground levels, and they eventually took out about seven million tons of ore from there.

Harold Lakes had a stroke before I came to the Emerald. He had been working so hard trying to get everything going. After that he lived in Nelson, but he occasionally came out for a visit.

When Harold's samples back in 1940 turned out to contain scheelite, he got an ultraviolet lamp, took a black blanket, and went searching for the ore. His wife was worried that he would be shot for a bear because he was crawling around under that blanket with his lamp. He would crawl down into the pits hand-dug in earlier times by prospectors looking for gold. The rock in the holes was mostly pyrrhotite, and when he shone his light on it under his blanket, the whole thing glowed and he found scheelite all over the place.

Because of the Korean conflict the Canadian government had repurchased the Emerald Tungsten from Canex in 1951. Canex had no facilities for assaying the tungsten from the Dodger and had to send the samples out to commercial assayers. Meanwhile, they got pretty good at estimating the grades, but it was not long before they were being overly optimistic in what they saw. Clive Ball wound up calculating reserves for the Dodger that were unrealistically high. Meanwhile, Simpson negotiated a contract with the US government for $63.00 per unit (20 lbs). Unfortunately, they could not find enough ore in the Dodger to meet their contract, so they were forced to buy back the Emerald Tungsten from the Canadian government at twice what they had paid for it. Everyone had to take some blame for this fiasco, but Clive Ball was the guy who took most of it. Nevertheless, in the end they made a lot of money from the tungsten. I have heard one figure of $25 million out of the tungsten. The lead-zinc made good money, but they benefited much more from the tungsten.

Corey Reyden at the Canex Reunion. Courtesy of Al Nord.

When the price of lead and zinc dropped in 1952, Gordon wanted to close the mine, but Simpson said, "You've got a good crew. If you lose them it will cost you more in the long run than to operate for three months at a loss." So they stockpiled the zinc for a while.

Corey Reyden had been in charge of the tramline and was the only man who could do a long cable splice. When the price of lead and zinc was up in 1950, they kept trying to feed more and more ore into the tram line. The cable finally broke and the buckets on their way up came roaring back down and tore out a bunch of the towers. The company had no drawings for those towers, so they had to take the twisted steel up to the shop and straighten it as best they could, take measurements off it, then cut new steel. Corey worked day

JEWEL OF THE KOOTENAYS / 107

and night to get the tramline rebuilt. Then the superintendent, who was a hard-nosed man, demoted Corey. I guess he needed a scapegoat.

Andre Orbeliani was a White Russian prince, some of whose mail was still addressed to "Prince Orbeliani." His mother was a painter who lived until she was 102. His wife Irene's mother was Mrs. Rode, who also painted. We were very friendly with them. They were interesting people, always doing something. When they celebrated the mother's 100th birthday, the mother played the piano for the whole party. Both Andre and his wife died only a couple of years ago, so they must both have been well up in years. They had no children.

Andre had degrees in both mechanical and in electrical engineering. Since Nick Smortchevsky also had an electrical engineer's degree, Andre worked in the mechanical end of things and designed a tremendous amount of the fabrication. They eventually retired to Nelson.

Andre was a terrible driver, and when he eventually was unable to pass his driver's exam, he taught Irene to drive. With him for a teacher, she wasn't any better.

When Andre retired, he wanted to write, so he wrote a little book called "The Scarlet Lady," which he wanted everyone to read. Someone said to him, "Andre, you have had a wealth of experience including work in Africa chasing diamonds." So he wrote one that was more readable on his diamond experiences.

Nick Smortchevsky's wife was a real character. She could swear like a trooper, but Nick was a real gentleman. He spilled some wine once and was so apologetic about spilling it, even though I told him it didn't matter at all.

At UBC, they got Gwen Gordon to participate in a geriatric study on how older minds operate. The doctor said to her, "Mrs. Gordon, you have the mind of a twenty-year-old." She replied in a huff, "I'm not that immature!"

In the Invincible mine, Clive Ball was having trouble. The drillers had gotten a drill bit stuck in the hole. They were trying to recover the bit and the core. There was an accountant at the mine who started bugging Clive about it. He said, "It's costing more trying to recover the bit than the bit is worth." Clive didn't tell him that he was trying to save the hole, not the bit. But he got fed up and left within a few months. He had been trying to ascertain whether or not the tungsten ore was overturned there, as it was in much of the Emerald mine. The hole had gone through the skarn band and back into the limestone. If he had continued the hole, there was a good chance that he would have encountered another occurrence of the tungsten ore, for it was exactly the same trough structure as in the Emerald mine. That would have made him a hero.

Archie McCutcheon was the mine superintendent for a while but left and went to the uranium fields at Blind River, Ontario. Later he and one or two others formed the Canadian Mine Services company and did a lot of work in the industry. Once at the mine, when the clouds covered the valley below, his young son, looking out over their expanse asked, "Can I walk on those clouds?"

At the main camp there were two gals who came up and stayed in one of the men's bunkhouses for a week. The miners were packing meals over to them from the cook house until Colonel Perry finally caught on. He was some upset. "They're not even paying rent!" he exclaimed. On another occasion Al Horton was staying in the staff house until a house would be ready for him. He sneaked his wife into the staff house and brought her meals from the cookhouse. After a couple of weeks someone caught on and let Colonel Perry know. Perry promptly chased her out, so she had to go back to Nelson until the house was ready for them.

Horton stayed around for quite a while until he went into exploration. He was asked to go to the Philippines to look after their property there. It took 15 years from when they took over the property until it finally became a mine. Horton, who was a sociable fellow, partied around with government people as part of his job, but became an alcoholic in the process. When he came back to Canada he wanted to make a world tour of it by ship. He said, "When I came from England to Canada, I tried to drink the ship dry, but found it was impossible." He was in the main office on Burrard for a while. I went down in the elevator and tried to strike up a conversation with his wife who was waiting in the lobby. She was waiting for Al to come down at 10:00 because that's when the bar opened. She wanted to nail him before he got to the bar. Later Al was having DTs, and Joe Adie put him in the hospital where the doctor told him, "If you don't quit drinking, you'll be dead in a year." Al quit and never drank again. After that he was managing the property in Alaska for a while and then he was in the San Francisco office until it closed, after which I think he retired. He was such an interesting fellow.

I went over to have my first look at Craigmont in 1957. After looking at the drill cores I phoned Doug Little in Vancouver, who was in charge of exploration, and told him, "This property has been described as a phase of a batholith, because they were trying to relate it to the Highland Valley copper." I said, "This is skarn. This isn't part of the batholith. It's high-grade copper." That was on October 10, 1957. Sarco, who had the Bethlehem Copper property, had a first right of refusal on Craigmont that expired on November 4 of that year. I talked to one of their engineers, who told me, "Oh, that's not a porphyry copper. It's not going to be big. It will never make an open pit." They were having problems with other prospects at the time, so they turned it down. Two days later, Placer had a deal on it.

I had a public company for 25 years—"Better Resources." When we ran out of money, I had the opportunity to turn the company over to somebody else. It is now called Blue Rock Resources. I am chairman of the board, but I don't have any real capacity except stirring up some old projects, for they retained some of the properties that I had acquired.

It sometimes seems as though there was never a dull moment at the mine. I remember when Steve Zuk was the mechanic looking after the underground shop in the Jersey. He and Jack Robinson disagreed about some of the equipment. Later, during a party at our house, the two of them got into an argument in the back yard. When they came in for coffee and snacks, they were still going at it. As they were coming in the door, Jack turned to Steve and said, "Fuck you, and the boat that brought you over." Life never got dull up there.

The curling rink had two sheets of natural ice and was built with volunteer labour. They had a party down there one time and had ploughed a trail down to it, but you couldn't drive to it. Everybody got pretty loaded. The door had a two-by-four for a sill, and this fellow who was leaving tripped over it and went sprawling on his hands and knees. He tried to get up but couldn't make it, so he was crawling up the trail on all fours when Gunnar Adolphson looked out and saw him. "My, my," he exclaimed, "The bears are out early this year."

At the swimming pool they put copper sulphate into the water to prevent algae from forming. Once when they overtreated the pool, the kids with blond hair who spent too much time in the water got bright green hair from it.

Jack Robinson was always a talker. He'd complain, "I've got so much work to do" and I would say, "Jack, if you stopped talking, you would get more work done." I saw him last about three months ago, before he moved into a nursing home.

When Stan Hill got married, Hub Maxwell helped them move into their house. He used the opportunity to tie a cowbell to the bed springs. Ten years later, when Stan moved out, the bell was still there, but the clapper had been taped over to silence it.

Terry Sanford was an electrician who lived in one of the cracker boxes in our end. He was pestered by a bear knocking over his garbage can, so he hooked the can up to his 110-volt power supply. The bear came and stuck its head into the garbage can, then belted it, BAMM, and the can went flying through the air. Terry said, "I sure wish I had hooked it up to 220."

JAMES GRAY

Mr. Gray is another miner I have known for a long time, but not well. I visited him in Ymir, where he has lived since the fifties. He was at the mine from 1951 until 1972.

My dad began mining underground in Wales when he was 11 years old. All the kids did that then. I came to Canada in 1927 on the *Montcalm*, a ship that later carried troops to Europe during World War II.

I first worked in a coal mine at Drumheller, where I drove a pony underground. I was a bit nervous at first. Altogether, I spent 15 years in coal mines.

James and Theresa Gray of Ymir, BC, 2006.

I married Theresa in Blairmore 61 years ago. She's born there, and we met in a curling/skating rink.

I was at the Emerald for over twenty years, and I remember breaking rock on the grizzly when Bill Toomey was a foreman there. I pretty much did every kind of mining job they had, including time spent scaling from a Giraffe that could reach 50 feet and later on, one that reached 85 feet. I have lived in Ymir all the years I have been in the West Kootenays and have always commuted to work from there.

I well remember George Soja, Bill Peligren, and John Kozar from both the Emerald and Remac. I recall that John was always a heavy gambler.

After the Emerald closed in 1973, I spent five years at HB mine before retiring.

I liked mining. It's all I ever did.

CAROL AND GILBERT WILSON

I met Carol and Gilbert for the first time when I interviewed them in their beautiful Panabode home in Salmo in January 2007. This was a case of a mountain boy marrying a mountain girl, for they both grew up at Canex. The only other such case that I am aware of was that of Pamela Gordon and Dennis Milburn. Carol and Gilbert both lived at the mine from 1951 to 1960.

Gilbert led off about his early years.

I was born in Manitoba and moved to a ranch at Oatscott on the Arrow Lakes (near Needles) at age six. There my parents were bought out by a logging company shortly before BC Hydro expropriated the area for a dam and? reservoir.

In May, 1951, when I was nine, my dad, Floyd, moved up to Canex, where he became a security guard. We moved to the upper main camp above the office, below the Dodger 44. These were the first houses people occupied at the mine. Then the company moved us out of there and down to the Jersey townsite in about 1955, because they were blasting in the open pit below us, and some rocks were landing too close to the houses.

I took the last two months of grade 3 as well as grades 4, 5, and 6 in the school at the mine. We then bussed to Salmo for the higher grades and I graduated there. After graduation I worked briefly at the mine before joining the army.

Carol then chirped in with her memories.

In school, I was two years behind Gilbert and never did catch up to him. We began dating during our school years in Salmo and were married on January 7, 1964, when I was 18 and he was in the Canadian army at Kingston, Ontario.

I was born in Vancouver and was a year old when we moved to Blairmore, Alberta, where my dad worked in the coal mine as a welder. When I was eleven in 1955, the family moved up to Canex, where Dad got a job as a welder. I was there for only grade 6 before bussing to school down in Salmo.

I didn't graduate. I was in my second year in grade 11, when Gil came home for a visit and decided I was going with him to Ontario. So I quit

The Wilson home at the mine. Courtesy of Carol & Gilbert Wilson.

112 / JEWEL OF THE KOOTENAYS

school in June and went with him. I lived with a family there and looked after their children while they worked. So I got my room and board on top of my wages. Gil and I came back to Salmo the following January and got married on the 7th. His father was still working at Canex. After we were married, we went back to Kingston again and stayed there until Gil got out of the army the following year.

My first son was born in 1967, so I did not go to work at the mine but was a full-time housewife there. I have two sons, Ryan and David. Ryan now lives in Winlaw and he has three boys. David is an environmental engineer in Ottawa and he has two girls.

Gilbert then continued his story:

After leaving the army in late 1965, I came back to Salmo and started working for Al Nord down on the conveyors shortly after Christmas or early in January. I was there for only about three months before I got the staff job in the warehouse I had applied for, but I still got lots of time on the "muck sticks" around the conveyors.

I started in the warehouse just below the Dodger 42 *[there were three different warehouses at that time]* and worked there until 1969, when the company transferred me to the Endako mine near Fraser Lake, where we stayed for a couple of years. Canex then asked me to come back to Salmo to help shut down the mine there and look after the

No. 2 conveyor. Courtesy of Al Nord.

inventory. I stayed there until the day after the auction was over and everything was cleaned up.

After the Emerald closed, I was transferred back to Endako for a year or two, but I was bounced around a lot, and I decided I didn't like the north that much either. Our two children were just getting into school and I wanted to be settled where they could do their schooling without interruption. So we moved back to the West Kootenays and lived near Castlegar in a little place called Raspberry. We stayed there for thirteen years until my kids finished high school.

I remember that fishing in Lost Creek was a popular past time for the boys and some girls. It was some of the best fishing I've ever had. It seemed like no matter how many we caught, there were still lots left. We never ran out of them.

The children at the mine played together in a way that I have never seen since. They just seemed to get on so well together up there, playing all those funny little games we invented, like "kick the can," touch football, and scrub softball. We made ourselves a skating rink and cleaned the snow off it. The rink had no walls, but it did have some dirt that had been piled up when they levelled the area. We lost few pucks because we were not strong enough to lift them when we shot. Many of our hockey sticks were homemade as well. I can remember that every kid up to age sixteen would come down to the schoolyard in the evening to play ball or kick the can until we absolutely had to go home and our parents were starting to yell at us.

We had lots of bears because of the garbage dump. The game people would trap a few and move them up the valley, but with the prevailing winds blowing the garbage smells up the valley, it would not be long before the bears were back. We had from five to eight of them roaming around the townsite all the time. We used to go out and chuck rocks at them. They weren't considered to be scary back then. If a bear became a problem, Al Nord would go out there with his big rifle and shoot it. I do remember the cinnamon-coloured bear that he shot. Until you *[L. Jacobsen]* told us, we didn't know that he still had its hide hanging on his wall. I can never remember a really bad bear scare up there, but it was common for a bear to come up on your steps and try to get into the house. Mostly if you yelled at them, they would take off. They never did become a real problem for anybody, because if it looked like they might, they were quickly shot.

We had lots of dances at the mine, but it seems to me it was mostly staff dances. We had two community halls. We first had one down at the Jersey, and when that burned down, we used the one at the tungsten townsite *[previously the main office]*. It had a library in it as well as the dance floor. Much later, after we came back from the army, they built another hall in the Jersey townsite right where the old one had once stood. That was shortly before the mine shut down. We were still finishing it off when we shut down. When we first moved up there, the staff did not mix much with the hourly paid people, but the last ten years, this was no longer the case.

I remember one kid, who got drunk and slid down the hill, but could not make it back up again. Sixteen of us finally managed to haul him back to the top.

I think the boys were more rambunctious than the girls. The staff used to have parties in the curling rink after the Tungsten hall had been abandoned and before the new community hall was built. The rink was sort of the get together place for them. They'd have parties there, whether they curled or not. Anyway, the following day some of us teenage boys would sneak down there. We'd get in through a window that they had not fixed, before anyone came to clean up. There we'd find part bottles of liquor (some would be almost full) that we would take out and stash in the snow. Occasionally it got cold enough that a bottle of wine would freeze and break before we got to drink it but the hard stuff never did.

Carol cut in with a memory of curling:

That's where I learned to curl. We did not have artificial ice and it was often covered in water when the weather got warm. During bonspiels, having a crowd of people in there did not help either. Sometimes the water on the ice seemed like an inch deep. That made it impossible to get a rock over the hog line. We did not have special curling shoes like everyone does now. We just used our ordinary footwear. I still don't use my broom for balance when I throw my rock.

Gilbert continued with stories from his school days:

We used to get so much snow there that we would have to go down and clean it off the curling rink roof. As soon as it got up to three feet or better (the rink had a flat roof), we would get up and clean it off. Sometimes we had to do that three or four times in a winter. We'd get a whole raft of people including teenagers up to help out. After it was shovelled off, we liked to jump off the roof down into it. You'd have to keep your hands up so that the others could reach down and pull you out if you went down too deep into it. You would be stuck solid. You couldn't move. You'd go down into it just like an arrow.

I remember one time when I was in grade five or six. We were looking for something to do at noon hour. It had been really dry for quite a while. It was probably early in the fall and the school grounds were really dusty. For some reason or other, we had a whole bunch of burlap sacks that we had been using to mark our bases on the ball field. We decided to play a game where we dragged these burlap sacks around the school grounds and it raised a huge cloud of dust. It was spotted by somebody and they radioed the fire department and the whole mine got shut down and people came rushing down thinking the school was on fire. They even had the miners come out from underground. The staff came down from the main office too. They all came to put out the fire. Did we ever get heck for that? I think some of us older ones may have gotten a few whacks on our hands. We had just been having fun. That was all.

Carol interjected:

We kinda made our own fun. I don't remember the girls doing anything bad like that, but perhaps it is due to a selective memory. The only thing I remember is one time at Halloween we were all out and one of the girls with me decided that she was going to soap a window in the commissary. So we boosted her up to soap the window and she broke it. Her arm went right through the glass. She was really lucky not to cut herself, but we all felt so bad because she had broken the window and we were worried about her but she turned out to be okay. I don't think I will ever forget that.

Gilbert finished off the interview with:

I don't remember the girls and boys being segregated at all. They all played together. We used to play a game called "British bulldog." You'd have to get from point "A" to point "B" and you'd have to bull your way through a line of kids. It was a little like rugby without a ball. Once we got to high school the boys did not play with the girls anymore.

Kick the can was a game played by opposing teams. The object was to prevent an opponent from entering your area and kicking the can. You would begin by setting the can down. You'd all disperse and hide. The first person to get back to kick the can would win. You wanted to hide nearby without being found. I remember the big thrill of being the first person to get back and kick it. We played it a lot in a back alley.

In the wintertime we would of course have snowball fights. They would last until a person got hit in the head. Then someone would run in and tell the teacher, who would be obliged to come out and stop it. We had one nasty little kid, who would bring frozen snow balls to school. There was another kid who was a bit wacky. He tried to catch any cat he knew about and hang it. He got away with that for ages before someone finally figured out who was killing the cats.

JUDY WAKEFIELD, NÉE STEVENS

Judy and her husband live in Vernon, where I interviewed her. Her father, in addition to his other work, was also the fire chief at the mine. Is it a coincidence that Judy is married to a fire chief? The family was at the mine from 1951 until 1969, when her father retired.

My dad, Ralph Stevens, was in Vancouver when the Depression hit and shut down his job. That's when he went up to Anyox to work. Mother grew up in Victoria. She and Dad were married in Anyox, on the Portland Canal near Stewart. The place burnt to the ground shortly thereafter. A book titled, "The Town that Never Was" was written about the place. We lived at Stewart, BC, from 1945 to 1948.

When we left Stewart, it took us three days by freighter to get to Vancouver. We crossed the border and came to the Metaline Falls by Greyhound bus where Dad met us and picked us up with Taylor's taxi.

We lived at Sheep Creek while Dad worked at Remac for a year before going to the Emerald. We stayed put, and I was halfway through grade three when Dad got a house and we moved up to the mine at Christmas of 1951 (I was born in 1943). The Harold Lakes School was still only one room then, but they added a second room later.

Our house wasn't quite finished yet and we had to climb a ladder from the main floor up to the bedrooms until the carpenters had finished it off.

Dad was a machinist by trade. He helped install mills and stuff like that. I remember him walking the long conveyor belt—the longest one in the world

Judy Wakefield, daughter of Ralph Stevens, who in addition to his other duties, was the fire chief at the mine. Judy lives in Vernon with husband Mark who is also a fireman. 2007 photo.

Dean, Cory, Judy, and Mark Wakefield. Courtesy of Judy Wakefield.

The Wakefield backyard enclosed by rock walls and complete with a stone barbecue. Courtesy of Judy Wakefield.

The road down the mountain. You don't want to slide off the road here. Courtesy of Judy Wakefield.

then. At Canex he also looked after the surface crew. On Saturdays I would sometimes go around with him and look into the portals. His machine shop was just above our house.

Across the street from the school was the big staff house where a lot of single people stayed. They used to have community dances there too. One night after a dance, I awoke and it was like daylight outside because the place was on fire. My dad, who was part of the fire crew, had to fight the fire in his dance clothes. After it burned to the ground, they just cleared the area and made it into a skating rink.

One thing about Canex was that they did a lot for their people. They used to bring a dance teacher up to the mine every Saturday morning from Nelson to teach us ballet, tap dance, and tumbling. They also showed movies in the cookhouse on Friday evenings.

We had noticed all this water running down the

mountain side from the compressors, so someone asked, "Can we have a swimming pool?" They said, "You guys build it and we will supply the materials." So everybody got together and built this huge pool with heated water from the compressors. The company hired lifeguards, who also gave us swimming lessons.

Once when we were driving back home from Vancouver through the States, Mom got violently ill near Colville. We took her to the nearest hospital, where they performed emergency surgery on her and she stayed there for two or three weeks. Dad was wondering how he was going to pay for all those medical bills, but Canex paid them all. I think Mom had a ruptured cyst. My twin sister and I were about 10 and my brother would have been about 17.

When Dad retired in 1969, he moved over to Wilson's Landing on Okanagan Lake where he had a lot. Because Dad was on staff, he had worked countless hours of overtime without pay. In appreciation, Canex gave him two Panabodes. They trucked them to the bottom of the hill, and he took delivery of them there.

We had a community Sunday school there every Sunday and community church service every Wednesday night. My mom taught Sunday school, Girl Guides, and Brownies. They also held Halloween and Christmas parties for the kids. My sister, who is five years older than me, used to work in the office during the summers.

I left the Emerald in 1961 to take practical nursing in Vancouver. My parents left a few years later when Dad retired, but then he was asked to go down to McDermott, Nevada, to look after installing a big mill there.

My first teacher was Doris Boyce from Nelson and Miss Jake from Lumby. She is married to Clausen and lives here in Vernon. The second set was Mr. and Mrs. Phillips.

Sunday school classes in the rec centre. Courtesy of Winnie Zuk.

JEWEL OF THE KOOTENAYS / 119

While my brother was in high school during the uprising of the Sons of Freedom, every school had an armed guard. My brother would come home from school, get his shot gun and sit with it on the school steps. You couldn't buy a pocket watch then without registering it. It was during the time when the train rails were being bombed between Nelson and Krestova.

We had to learn to live with the bears and they would sometimes be on our back porch. Sometimes coming home after babysitting we used to run pretty fast down the trails because we knew the bears were out.

My brother went to the mechanic school in Nanaimo, did his practicum in Cranbrook, and then came back to the mine and worked as an underground diesel mechanic. He worked there for a couple of years but coughed constantly, so his doctor said to get out of there. After that, he worked for International Tractor for many years as a mobile mechanic, going out in the bush to repair equipment.

One day at the mine Dad brought home a letter that had just arrived at the office. It was one he had written to my mother 10 years earlier while we were still in Stewart in which he described the Kootenays.

My mom lived to the age of 67, and Dad died at Kelowna in 1996.

I came to Vernon in January, 1965, married Mark Wakefield two years later. We had two sons, Cory and Dean and we all still live here in Vernon.

I found that the Emerald was a great place to grow up and have never regretted spending my childhood there.

CAROL MCLEAN, NÉE ANDERSON

Carol is the daughter of the late Clarence Anderson, who was in charge of all surface transportation at the mine. The family lived at the mine from 1951 to 1968. Carol and her husband now live near their children on acreage a few kilometres north of Salmo and provided me with excellent aerial colour photos of the Emerald.

My earliest memories are starting school, the great outdoors, swimming, hiking, and lots of snow. We all tobogganed until we were so cold we could hardly make it back home. We had a great skating rink, which was nice. The kids shovelled the snow off it when they wanted to skate. Every year we had a lifeguard—quite often it was someone who had grown up in the community, gone off to college, and come back to teach during the summer. My brother Arnie taught some lessons one year. We were totally spoiled, for we had the only heated swimming pool in the region. *[When I asked if it caused them to look down their noses at other people, Carol's husband Dennis interjected, "That wasn't hard to do when you lived on the top of the mountain."]* I don't think we really knew at that time that there was another world out there. We went to Nelson once a month for groceries. That was our big outing.

I think the most memorable part of Canex for most of us was the swimming pool and that was because we all took swimming lessons. As soon as the pool opened in the afternoon, we stayed there until supper and if we got to go back after six to eight o'clock in the summer, that was great.

There was just a single pay phone at the mine for the entire community. We had other phones with which we could call others in the community, but for calling Salmo, we had to use the pay phone, and it was by the minute. So we used to call whoever we had in mind, then hang up and let them call us back because they did not have to pay.

We started out in a Panabode, but there were six kids, so we moved into a duplex where we got both ends of it. One end that had been a kitchen was turned into a bedroom where my brothers, Arnie and Kenny slept. I had the big bedroom, which I shared with my sister Joyce. My brother Roy slept off in a little bedroom. Keith slept in the next one, and my mom and dad had the remaining master bedroom. We pretty much needed the whole duplex, but we did live in a Panabode for quite a few years.

Carol, née Anderson, and husband Dennis McLean.

Carol's brothers, Arnie & Ken Anderson. Courtesy of Ernie Stenzel.

Few people tried to garden because the winters were so long, but I remember Clare Vayro did have a small vegetable garden, because some of us would go "coon hunting" at night and pick a carrot or two from it. There may have been a dozen people up there who grew gardens of sorts.

My husband Dennis's strongest memory was of the wintertime and trying to get up to the mine to court me. That was his biggest challenge—getting there. The winters on the hill could be brutal.

DENNIS AND PAMELA MILBURN, NÉE GORDON

I twice met with the Milburns in a crowded Denny's Restaurant near Langley, BC. Pamela grew up at the mine and arrived with her parents in 1951. Her father, Gerry Gordon, was the mine manager during the summer I worked in the Dodger 44 Tunnel in 1952.

Dennis led off the conversation.

I was born on January 20, 1940, and wound up as both a geologist and an accountant. My father, Bill Milburn, was a mechanic at the tungsten mill. His mother was a member of the pioneer Grutchfield family that came to the Salmo area in the late 1800s. Many Salmo people are related to me.

I remember working during the summer and saving enough to keep me in university all winter. The mine management had a policy of hiring the children of employees; so many students were able to continue their college and university education courtesy of the Emerald Mine. Tuition was $400 a year when I went to UBC in 1958. Later, when they developed the Craigmont mine, a lot of the students went over there to work too. Canex was good for a lot of the students.

I remember that Canex had a research facility at the Dodger concentrator, where they processed and assayed all the samples from the Endako mine, and later from the Marcopper mine in the Philippines as well. I was the flunky working with all the testing of that ore at our facility. We had a small ball mill, miniature flotation cells, and all

Wedding photo: left to right: Bill Milburn (mechanic), Mrs. Lu Milburn, Dennis Milburn, Pam Milburn née Gordon), Mrs. Gordon, Gerald Gordon (mine manager). 1964 photo. Courtesy of Pamela and Dennis Milburn.

those sorts of things to do with milling the ore. Bert Wilson was the chemist in charge of that research lab. It was the forerunner of the one we later had in Vancouver.

Dennis's wife Pamela then told me about her time at the mine.

I am the daughter of Gerald Andrew (Gerry) Gordon, who was the general manager at Canex from 1952 to 1958 before moving to head office in Vancouver. My father worked at the Cariboo Gold Quartz at Wells, BC, before coming to the Emerald in 1949 as a mine superintendent. We first lived in a duplex, in the Emerald townsite, which we heated with a wood stove. We also had to use an "outhouse" until the company provided the houses with inside plumbing.

From the Emerald we moved to the Jersey townsite and then into the "manager's residence" at Main Camp after Dad became the mine manager. My mother's maiden name was Kelly, and both her parents came from Ontario. I remember my father telling us that the Emerald Mine was designated as a "safe house" for the BC government in case of war. They would be able to stay underground there and have good air and water.

I have two brothers: Terry, who is a geophysical engineer, and Steve, who is a chemical engineer. Terry and Dennis were close friends, and Terry was probably somewhat responsible for Dennis marrying his sister.

I worked at UBC in Education for a number of years and have three related degrees. While there, I wrote various books for BC schools, including ones for social studies, art, and Native history. School teachers all over BC now use the art book I wrote. I am presently writing a history of my family, that is, whenever I can take time out of a hectic schedule of renovating our home in Langley.

I remember that when Harold Copley was away on weekends, he hired me, a thirteen-year-old, to do his second job of cleaning the main office and mine dry. I made a dollar or two for each job and thought the pay was good.

Bernard Feeney, a relative of Dennis's, supplied fresh milk to the mine from the farm he owned at the base of Iron Mountain that is now called Stagleap. He sometimes chased trespassers off his land with the aid of a firearm. When people complained to the local Mountie, he reputedly said, "If you want to stay alive, stay off Feeney's farm."

As a child on the hill, I remember a host of activities that included chasing bears, fishing, hiking and playing in the woods, sledding, skating, playing softball, and downhill skiing without the benefit of rope tows. I was involved with the Brownies, Cubs, Guides and Scouts organizations and took mine safety courses. To earn money, I collected beer bottles and worked as a cleaner in the mine dry. The school, community hall, swimming pool, curling rink, and the commissary were all located at the Jersey townsite.

COLIN BROWN

Colin was at Canex from 1951 until 1968 and was its second last manager (from 1964 to 1968, when Ed Lawrence took over). After a year and a half at Craigmont, he joined the "Salmo Mafia" at the head office in Vancouver, where he stayed until his retirement.

I started at the Emerald Tungsten Mine in 1951, as a junior engineer. On arriving, I shared a room with an assayer named Bert Wilson in the staff bunkhouse above the cookhouse. The kitchen staff stayed on one side of the top floor, and the engineering and supervisory staff on the other.

Unlike a lot of the engineers, I never did work as a miner. I worked initially as an engineer for six months before going into the mine as a shift boss. One of my first jobs as an engineer was laying out a raise to be used as an ore-pass from the Emerald 3,800 level to the 4,200 level. The miners were getting close, when the mine superintendent, Hub Maxwell, asked me if I was sure it would break through in the right place. I marked an X on the wall and said that that was where it would come through. The following day the first drill hole came through in the middle of the X. Hub was impressed.

Colin Brown, 2007.

J. D. Simpson, president of Placer (the owner of Canex), came up to the mine on periodic visits, particularly when the Tungsten operation was getting close to production. I remember during my early days at the mine I was really gung-ho to do a good job, and often worked late in the evenings. Mr. Simpson would often come down to discuss ore reserves with the chief geologist, Quint Whishaw. I guess I made a favourable impression on him, as later on I was given jobs all over the property. That made me familiar with the whole operation. Years later I felt that they were possibly grooming me for greater responsibility.

In 1953 my future wife Eileen came to work in the accounting office at the main camp. She had trained as an IBM keypunch operator at the Britannia Mine and replied to an ad for a similar position at the Canex operations in Salmo. She worked in the payroll office with Bill McDermott and lived in the "Girls' House" at main camp with Binky Wragge, the manager's secretary, Beth Chalmers, an assayer, Doris Boyes, a schoolteacher, and Betty Sanderson, who worked in the warehouse. We met at a dance in the community hall, the one-time cookhouse for the Emerald operations.

Cracker box houses, mid-1950s. Courtesy of Colin Brown.

After we were married, we moved into a little frame house 20 feet wide by 24 feet long, and brand new. It was actually half of a duplex, and there was a crack at the bottom of the wall separating the two units. We had no secrets. You could hear everything that was said on the other side, but we had good neighbours. Soon after we moved in, our neighbours acquired a television set, so we, having to keep up, got one too. Most homes at the mine had TVs, and we could receive two channels from Spokane. I believe we had more TVs per capita than most places in BC, including the Lower Mainland, but they were black and white of course. The following year, after our first child was born, we moved into a frame house of about 600 square feet. It seemed like Paradise.

While at Salmo, I was moved all over the place. I was the mine engineer at both the Emerald and the Jersey, as well as a shift boss at both operations. I was also surface foreman for several months. While I was the shift boss at the Jersey trackless operations, we first mined the top layer and then came down through the ore body in benches. We used to have friendly competitions to see which crew could get the most ore out, in a shift. The shifter opposite me liked to make sure that it was my crew who drilled and blasted the bench, and then he could muck it out. He always crowed about how many more loads his crew got than we did. One day we put two jumbos on the bench, drilled and shot it out before lunch, and then mucked it out after lunch. We could only do this two or three times before they caught on. Miners are very competitive. They all want to be the top dog, and get the largest bonus.

Gerry Gordon was my first manager at Salmo, and Doug Little was his assistant. Both were good at their jobs, and great people as well. Later, Graham Walkey was the assistant manager for two or three years. He took me under his wing and tried to keep me on the straight

The mine manager's house, mid-1950s. Courtesy of Colin Brown.

and narrow. When he went to the Stanleigh mine in Elliot Lake, he asked me if I would like to come and work for him there. Hub Maxwell at the Denison Mine in Elliot Lake also asked me if I would like to work for him. It was 1957 and the ore reserves at Salmo looked rather bleak, so it seemed to be only a matter of time before the mines would close. I decided to accept Graham's offer and resigned. Gerry Gordon spent most of an afternoon trying to convince me to stay. I told him that I couldn't see too much life left at Canex, and asked if he could promise me that I would have a job for the next few years. He said that nobody could make that sort of promise. I left as I believed there were many more career opportunities at Elliot Lake.

Because we had the first trackless mines in Canada, there were a lot of Canex people in Elliot Lake, which was largely trackless too. We therefore had many friends there. There was a company townsite at Stanleigh, but we lived in Elliot Lake and commuted about 15 miles each day.

In 1960 I received a phone call from Charlie McGowan, then mine manager at Canex, offering me a job as Mine Engineer at the Jersey operations. Since Elliot Lake was not the nicest place to live and work, I accepted. When we returned, it was like coming home. We first moved into another frame house, and then into one of the larger duplexes that had been converted into a single unit. There we had lots of room.

When I returned, the Craigmont mine was being readied for production. Bob Hallbauer, the mine superintendent at Salmo, went to Craigmont in that capacity. Jack Robinson took over as superintendent when he left. Tony Triggs was chief engineer at Salmo, but he also left for either Vancouver office or Craigmont, and I took over as chief engineer. At that time Bob Weber was the manager at Salmo.

In 1964 the Endako Mine was being readied for production, and Bob Weber went there as manager, so I became the mine manager at Salmo. I held that job until I was moved to Craigmont in 1968. I stayed there for a year and a half until I was transferred to the Vancouver Office where I stayed until I retired in 1982.

When we left Salmo, Ed Lawrence took over from me as mine manager. He has been up there again for some time with Sultan Minerals Inc., looking after a drilling program. Ed is a good man for the job, for while he was the manager, he did a lot of exploration in the Tungsten areas.

Other Salmo people who wound up in the Vancouver Office were Gerry Gordon, Doug Little, Charlie McGowan, Harold Steen, Tony Triggs, Doug Knight, Berte Wilson, Earl McLean, Nick Smortchevsky, Mel Olson, and I believe Blue Evans. Blue was not there very long, as he went on to teach at UBC and was later dean of mining there. Ex-Salmo employees in head office were often referred to as "The Salmo Mafia."

We have three sons, Rory, Howard, and Derek, who all began school at Canex after we came back from Elliot Lake. Rory and Howard got their first six years at the mine school, and Rory also went to the Salmo high school for one year.

There are always both good and bad things, but mostly one remembers the good ones. We had lots of bears at the mine, especially on garbage day. One day Nick Smortchevsky, the chief electrician, thought he saw a dog in his garbage He went out and swatted it with a broom. Imagine his surprise when it turned out to be a small bear.

We did have mine accidents, but I prefer to remember those with a good outcome. In the Gil Mosses affair, the men did a wonderful job in getting him out. Werner Trachsel, a neighbour of ours, was also trapped in an ore-pass, and it took them about 12 hours to free him. He was not badly injured, but the Slusher cable was between his knees, and he always had difficulty walking after that. He probably had some nerve damage.

As an aside, a few years ago on Father's Day I received a phone call from my son Rory, who said, "Dad, guess where I am. I'm on top of Iron Mountain." (He goes up to the mine site often, as he has many fond memories from his childhood there). I asked him, "Do you know how to get down?" He replied, "I've got a GPS." I thought, my gosh, if he's on top of Iron Mountain there are a lot of open holes he could fall into. Later he told me that the GPS showed the ore bin, and sure enough it led him right to it. However, all that is left of the townsite is the shell of the swimming pool.

THE RUSSIAN CONNECTION: SMORTCHEVSKY FAMILY

Nick Smortchevsky. Courtesy of Nick Smortchevsky Jr.

There were a number of "larger than life" characters who worked at Canex at one time or another, but some of the least likely ones had to be the two Russians, both of whom had fled Russia during the 1917 revolution, and both of whom led fantastic lives.

According to the CP Air ticket reproduced here, the Smortchevsky family left Shanghai on August 18, 1951, and flew to Vancouver, where they

Nick and Tania Smortchevsky. Courtesy of Nick Smortchevsky Jr.

Nick Smortchevsky Jr., B.Sc. (Eng.), came to Canada from China in 1951 with his parents and siblings. His father was the electrical superintendent at the mine. His parents fled Russia during the 1917 revolution. 2008 photo.

stayed for six months. Nick may have worked at the Nickel Plate mine at Hedley before moving up to the Emerald Mine, where he became the Electrical Superintendent.

At the Emerald Nick Smortchevsky was regarded as a genius. One feat that may have helped cement that impression was that he invented and installed an early warning system to detect approaching lightning storms. Lightning was a frequent event during the summer, and not only did it interfere with the electrical equipment, but it also posed a serious hazard for miners wiring electrical blasts. On one occasion it caused a premature detonation. Fortunately, no one was injured in that episode. Tony Triggs said to me, "Nick was the most intelligent engineer that I ever ran into. He did all the electrical design for Placer. He was also a mechanical engineer and went back to basics when he wanted to figure something out, for we had a lot to learn back then."

The Smortchevskys had two sons and a daughter. I was able to track down and meet with the older son, Nick Jr., while he was on a brief visit to his home in North Vancouver. Nick is presently spending most of his time in Ecuador, where he is the project manager for Aecon AG Constructores SA on the construction of an airport. He provided me with a few photos of the family and the following personal account:

NICK SMORTCHEVSKY JR.

I turned six while we first lived in Vancouver, and I started school at Henry Hudson before we moved to the mine site, where I finished my first six grades. I then attended the Salmo high school for the next four years and was in grade eleven when we moved back to Vancouver. After completing high school there I went on to UBC, where I got a civil engineering degree and went into the construction business.

Tania Smortchevsky. Courtesy of Nick Smortchevsky Jr.

The Smortchevsky children: Adrian, Nick Jr., and Natalie. Courtesy of Nick Smortchevsky Jr.

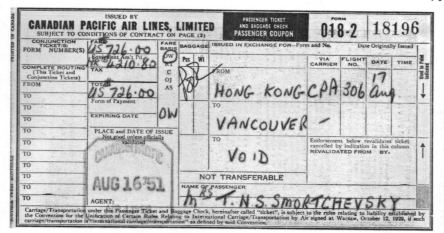
CP Air ticket. Courtesy of Nick Smortchevsky Jr.

Dorothy Hartland during mid-fifties. Courtesy of Lewis Hartland.

My parents were White Russians who escaped to Manchuria and later moved to China, where my father finished his education and got his degree in electrical and mechanical engineering at the Polytechnic Institute. They moved to Shanghai when they were in their early thirties, and we children were all born there.

I remember that my classmates at the Emerald were Colin Hartland, Elaine Stenzel, Gilbert Wilson, Clinton Peters, and Keith McLeod.

[I was able to speak briefly with Nick's sister, Natalie Olson, over the telephone. Here is what she had to say:]

JEWEL OF THE KOOTENAYS / 129

NATALIE OLSON, NÉE SMORTCHEVSKY

I don't think my father worked at Hedley. I think he worked for a machinist for about five months until he got his Professional Engineer status.

I remember my mother, Tania, experiencing quite a culture shock. In China she was accustomed to having servants do all the physical work. At Canex she had to learn to do it all herself.

We spoke Russian at home, but gradually switched to English once we children had learned to speak it. There was no multiculturalism in those days. People were told, "Learn English—that's the only way to get ahead." However, once my parents became old and enfeebled, they reverted to Russian. Both of my parents were also fluent in Chinese.

I began school at the mine the following September. I feel fortunate to have had committed teachers. I remember the mine society as being quite stratified—the staff wearing white hard hats to differentiate them from the hourly paid people. Later, when my husband and I lived at Cassiar, we found the same schism between salaried and hourly employees, but even more so. Mother didn't care about status, and Dot Hartland, who was a war bride and lived next door to us, became her closest friend.

Regardless of how the grown-ups got along, we children knew no class distinction and we all played well together. I found the mine a great place to grow up.

We stayed at the Emerald until 1961, when Dad moved into the head office of Placer Development, the parent company, in Vancouver. He stayed until he retired and passed away in June of 2005.

THE ORBELIANI FAMILY

Prince Andre Orbeliani fled Russia with his family at age 17, in 1917. After several moves they settled in Zagreb, Yugoslavia, where he finished his education with degrees in both mechanical and electrical engineering. He came to Canada after spending a number of years in Africa in what is now Zaire as well as in Belgium.

Smortchevsky and Orbeliani were both brilliant, but Andre's electrical knowledge did not match his mechanical. He became the mechanical superintendent and earned a reputation for the quality and elegance of the plans he drafted in that department at the mine.

Andre brought over not only his wife but his own mother as well. Both women were artists who painted, and Andre's mother demonstrated her prowess at the piano right up until she was 100 years old when she played at her own birthday celebration.

The following excerpts from news clippings and obituaries tell this family's fantastic stories.

Former Nelson artist feted on 100th birthday[9]

Mrs. Orbeliani is seen here on her 100th birthday as she chatted with Mrs. Edna Whiteley.

Mrs. M. Orbeliani cut her 100th birthday cake and entertained friends at the piano with a masterly rendition of The Blue Danube Waltz. *[Her birthday was actually on June 20th.]*

Mrs. Orbeliani, whose official birthday is June 20, was entertained by the women of St. Paul's-Trinity United Church Sunday following the morning church service.

This alert little centenarian is an accomplished artist and has had many showings of her work while she lived in Nelson. She has resided at Penticton for the past few years.

Flowers in the church were dedicated to her Sunday morning and she was presented with a rose corsage which was pinned on her becoming salmon pink coat dress.

She made a charming speech to the more than 100 gathered there including many from the other churches. Among the guests were Mr. and Mrs. John Erb of Vancouver.

Within the past two years she has had an art showing at Penticton and is now busy writing her memoirs. These should be fascinating for she was in the inner circle of the Russian royal family, during the reign of Czar Alexander III, his son Nicholas II, and Czarina Alexandra, Queen Victoria's granddaughter.

There was her flight from Russia during the revolution—to the Caucasus, Georgia, Constantinople, Yugoslavia, Belgium then to Canada in 1952, as a widow. Her husband had died in 1948. She came here to join her son Andrew. She taught school at St. Margaret's in Duncan from 1956 to 1958 and can speak five languages, French, Russian, Yugoslav, German and English.

Mrs. Orbeliani is returning to the rest home in Penticton where her 100th birthday is to be celebrated Thursday.

Mary Orbeliani. Courtesy of the Nelson Museum Archives.

---

9 Nelson Daily News, *June 18, 1974; Nelson Museum Archives.*

Mary Orbeliani dies[10]

Princess Mary Orbeliani, well known throughout BC for her beautiful water colors, died Saturday at Mount St. Francis, aged 103. She was born in Petersburg, Russia in 1874 to a family of Russian nobility. At 19 she was introduced to the Court of the late Empress Alexandra. She married Prince Alexis Orbeliani.

Following the Russian Revolution in 1917 she began a laborious life as a school teacher, first in Russia, then in Yugoslavia. Her husband became sick and disabled and she remained the provider for the family for many years. Life was especially hard for her during World War II.

When her husband died she came to Canada in 1952 to join her son and settled in Nelson. Even at that advanced age she wanted to be independent and earned her living with private lessons in French. At 82, this remarkable woman was teaching school at the Queen Margaret Boarding School for girls in Duncan, BC.

During the years she continued to paint masterful water colors and many of these may be seen in Nelson homes . . .

Andre Orbeliani's corrections to the above[11]

Mary was born in Wiceibudar, Germany, not Petersburg, Russia, but moved there shortly after her birth.

When her husband died (in 1946) she did not go to Canada immediately, but to Belgium to join her son, Andre. She moved to Canada in 1953 to be with him.

In Nelson and in Duncan, she did not work because she wanted to be independent, but because of economic necessity. In Nelson she was very popular, and had many students learning French.

The Passing of a Prince[12]
by Anne DeGrace, Daily News *Staff*

Long time Nelson resident Andre Orbeliani held the 2000 year old, *[200 years]* title of merit as a member of the Russian nobility. But according to many of his friends, neighbours and co-workers, he was a prince in ways that transcend title. Mr. Orbeliani died on October 9 of this year at the age of 99.

Born into the Georgian nobility in Russia at the turn of the last

---

10 Nelson Daily News, *December 5, 1977; Nelson Museum Archives*
11 Nelson Daily News; *Nelson Museum Archives*
12 Nelson Daily News, *October 13, 2000; Nelson Museum Archives.*

century, Andre's early childhood was one of title and wealth. His father was a high-ranking civil servant, employed first in Moscow, and later Siberia. Andre was an only child. For the week-long train journey from Moscow to Habarovsk, the family was given a private rail car. Vacations were spent in nearby Japan.

The First World War began, followed by the Russian Revolution. The family fled from Russia to Georgia as the occupation spread, rousting out nobility, as it went.

The family feared for their lives. Packing what they could, they trudged through high passes across the Caucasus Mountains, 17-year-old Andre accompanying his parents on foot, with a horse and cart to carry a few belongings.

Andre Orbeliani. Courtesy of the Nelson Museum Archives.

In Georgia, Andre started university. His father got a job while his mother, gifted in languages, taught. It was there that Andre's mother received word through the French consul that the train lines were about to be cut by the Bolshevik insurgents. The family fled again.

They landed in Constantinople, now Istanbul, and finally settled in Yugoslavia, where Andre attended the University of Zagreb, studying engineering. Later, he graduated from the Catholique Universite de Louvain with a double degree in electrical and mechanical engineering.

When the great depression followed, Andre had worked for just a year for a Belgian electrical consulting company before being laid off due to economic circumstances. Nevertheless, his employer thought so highly of him that he arranged for the young engineer to work for the diamond mines in Angola, Africa. Later, Andre worked managing a power plant in what was then the Belgian Congo, now Zaire.

Andre's wife of 75 years *[he was married to Irene, his second wife, from 1940 to 2000—not 75 years]*, Irene Orbeliani, remembers how well liked and respected her husband was at that time. Although she had not then met him, she did return to Africa with him after their marriage. She remembers a letter he received from one of the men who worked for him, expressing his appreciation.

"Andre loved it there. Then, there was no hatred. Black workmen would walk miles to get a job with him. Lots of white men were terrible to the Negroes, but Andre gave teaching to them," she said. "Black workmen wrote beautiful letters to Andre."

Long-time friend Marylee Banyard echoed this in a eulogy she delivered at Andre's funeral.

"In those days, in that mine, it was customary for white men, to hold most top positions, and African labourers did most of the dirty and often very dangerous work. When he told me of this situation, many years later, Andre was still distressed by this fact. He himself never asked an African to do a job he did not do himself, and so he was both loved and respected by the African staff."

Andre had been married in his last year of university, but his wife did not accompany him to Africa; eventually, they divorced. He worked in Africa for nine years before returning to Belgium in 1939 for six months leave. During his leave in Belgium, war was declared, and Andre was unable to return to Africa. He spent the war in Belgium

He met Irene in Belgrade, where he travelled with some difficulty across hostile German territory to visit his parents.

Irene had been attending university in Belgium, and a professor had encouraged to her visit his relatives in Yugoslavia: Andre's parents.

"He had a special aura," said Irene. "Andre was always very brilliant, and always a gentleman."

Andre and Irene were married on February 1, 1940 in Belgrade; they celebrated their diamond anniversary this year. They hoped to return to Africa after their wedding, but by then Germany had invaded Belgium and they were compelled to stay.

After the liberation of Belgium the couple left again for the Belgian Congo, returning two years later to work in Brussels. But after Africa, Belgium seemed too small, Europe itself dense and over-populated. They considered a move to Argentina, but when Canada opened its borders to all immigrants rather than just those from Commonwealth countries, they applied....

Andre was able to speak seven languages: Russian, French, English, German, Serbian, Croatian, and languages learned while in Africa and he had much experience adapting to different cultures and countries. Still, the leap from Europe's metric system to Canada's Imperial system of measurement was something he struggled with.

In 1951 Andre secured a job in Salmo at the Canadian Exploration mine of Placer. The couple lived there for nine years, at an elevation of 4,500 feet. Later, they moved to Merritt where Andre was design engineer at a new mine. After 12 years, Andre and Irene moved to Prince George to work at the Endako mine. In 1970 Andre returned to Salmo to redesign the Tungsten mill until his retirement at 71, when the couple moved to Nelson.

Andre was in demand throughout his career, highly respected for his expertise, and for his way with people; his late retirement was due to reluctance by the industry to let him go. Throughout these years, Andre and Irene made fast friendships, and the respect Andre garnered in his early days in Africa was repeated time and time again.

After his retirement, Andre had time to develop hobbies and interests, and these took him to the arts and other activities he had not previously had time to enjoy. He studied English with Don Wilson at DTUC, and the two remained friends. He wrote poetry, publishing a volume entitled Twenty-eight Grams of Poetry in the 1970s. He taught pottery, loved gardening, astronomy and Greek mythology, and was an excellent chess player. He enjoyed sailing and kayaking. He learned stained glass from Tam Shields and weaving from Irene.

"He had a computer in his head," said Irene of his weaving skill. "I have to look in a book, but he made his own design. He had a very good sense of colour."

Even with plenty to learn and keep him occupied, Andre still found retirement quiet. He volunteered to assist the City of Nelson.

She explained that during his final years at Jubilee Manor, during which time he suffered from Alzheimer's, Andre made friends with the nurses, who grew very fond of him. He was always courtly, kissing a hand as he expressed thanks. Once, she says, when he could no longer feed himself, he said of the nurses helping him, "three beautiful ladies feeding me, and I am very thankful."

With the passing of Andre Orbeliani, the world has lost a noble man.

Irene Orbeliani
October 3, 1911–March 9, 2005
passed away peacefully at 94 years at
Mount St. Francis on March 9, 2005.[13]

Irene was born in Russia to Leonide Zozouline, a naval architect, and to Barbara Rode, a painter. Fleeing the Russian Revolution of 1917, Irene and her mother travelled to Egypt and Yugoslavia, finally settling in Paris, France in 1920.

---

13 Nelson Daily News, *March 16, 2005.*

Irene Orbeliani.

Irene attended secondary school in Paris. In 1932, she went to Belgium to study at the University of Louvain, graduating with a degree in Political Economy. She then worked for the Ministry of Finance in Yugoslavia.

Irene married Prince Andre Orbeliani in 1939 *[1940]* in Belgrade. His work as a mining engineer took them to Belgium and the Belgian Congo. In 1951 they emigrated to Canada. They lived in Salmo and Merritt, moving to Nelson in 1972.

Upon arriving in Nelson Irene enrolled at the Kootenay School of Art, becoming a weaver thereafter.

Irene was a member of the Anglican Church and of the Weavers' Guild. She will be remembered as a good friend to many, a gracious hostess, a bon vivant, and a great patron of the arts. Irene was predeceased in death by her husband Andre in 1992 *[2000]*. She is survived by a cousin in Russia, a niece in Italy and a step-sister in Serbia . . .

HAROLD KETTLESON

Hal was an engineer at the mine. We were both there in 1952, but I did not meet him until years later, when we both worked in highway construction. I first knew of him because he had worked for Peter Kiewit Sons Company not long before I started with them as a "rock superintendent." Hal has lived in Vancouver for many years. The following material is based on conversations with Hal.

Hal Kettleson grew up in Vancouver and put himself through UBC by working as a miner at Bralorne from 1948 until he graduated in May 1951. He had also worked one summer in Alaska.

His wife Anna grew up in the Fraser Valley. She met Hal in Vancouver, while she was working in the admitting office at Grace Hospital and he was a student at UBC. She lived upstairs with her sister in the home of Hal's best friend, and that's where Hal met her. The two were married in Vancouver the Christmas before he graduated.

Before he came to the Emerald in 1952, Hal had worked in Yellowknife for Consolidated Mining and Smelting Company (CM&S, later called Cominco). He left Cominco because he was dissatisfied with both the house and the pay.

There were several people at the Emerald who greatly impressed Hal. One was Gerry Gordon, mine manager from 1952 to 1961. He came from Ontario and stayed with Placer for the rest of his career. Another was Colonel Perry, the chief financial officer at the mine. He was from England and had been a great rugby player. He wound up with osteomyelitis (a bone infection) and progressively over fifteen or twenty years lost first one leg, and eventually the other. But Hal remembers him as one of the outstanding people up there.

Another person to impress Hal was Hubert (Hub) B. Maxwell, "probably the most incredible man in the camp." He had only a grade-eight education but, in Hal's words, "a mind like a bear trap." He would ask questions and learned all he could from his co-workers. He became so proficient that he obtained his Professional Engineer status in Ontario because of his knowledge and work history—he had even become a mine manager. When he returned to BC and applied for membership in the Association of Professional Engineers here, Hal was only too happy to sponsor him. Everybody, whether ditch diggers or directors, liked and respected Hub. He was considered a good poker player too.

Simpson, the managing director of Placer, had the foresight to get a long-term General Services Administration (GSA) contract for the tungsten with the American government. This metal had always suffered from competition from Formosa and China, who had large deposits and cheap labour and could produce it at lower prices than anyone else. When the Korean War shut off that supply of tungsten, the GSA decided to stockpile it and considered Canex a great source for it. By negotiating a long-term contract, he ensured the orderly development of the tungsten ore bodies.

Hal also talked about Doug Little, who graduated in 1950. Apparently some of the staff thought that he should have become the president of Placer. He was a great and talented "down to earth" person but could not match the social skills of a Ross Duthie, who later became president, and his wife Betty. The paradox was that Doug and Ross had been buddies in school and Doug had encouraged Ross to come to work at Placer after he had already made his own mark there. Still, in Hal's words, "Doug did very well."

Another former colleague Kettleson mentioned was Harold Steane, the metallurgist at Canex. He was from Australia and a wonderful professional who also set a standard of excellence. He fine-tuned the milling process and made it work. When Placer started the Gibraltar mine, he decided on using rod mills and ball mills, while at the Lornex Mine at Ashcroft the new "free flow" mills were used. Some people criticized Steane for staying with the "tried and true," but he honed it to a fine degree, and while

Engineering office. 1950s. Courtesy of Marie Adie.

it took Lornex three years before its mill began to pay off, Harold Steane's process had paid off the whole plant at Gibraltar in the same time span. Steane contributed to the Emerald's success by installing a small roaster to process some of the tungsten concentrates; tungsten can be sold as scheelite, or it can be roasted and sold as an oxide, which fetches a better price. This innovation in the end contributed to extending the life of the Canex mine. When the Kettlesons moved into their Vancouver house, they found that Harold Steane was living only two or three blocks away.

The head of geology during Hal's time at the Emerald was Clive Ball, who had been a decorated pilot in the Australian Air Force. It wasn't known until years later that he had flown some terrifying flights from India to China. He was so polite that if he bumped into a filing cabinet, he'd say, "Excuse me." Years later, after a stroke, his demeanour changed and he became much more outgoing. He was well liked, but some people liked to joke about him.

Other contemporaries of Hal's at the Emerald were Tony Triggs, an engineer who went on to have a good career with Placer; Al Horton, a geologist who went to the Philippines and helped start Marcopper; and Cliff Rennie, also a geologist, who moved to and stayed at the Vancouver office.

Hal also talked about the event that helped initiate the lead–zinc operation—finding the outcropping of the main ore body. It had generally been thought that the lead-zinc occurred in "chutes" where the old mine was. What had not been realized at first, however, was that it was an ore body like an arm, which "fingered" out to surface (each finger being four to five feet thick). The main ore body was down below and was up to 50 feet thick.

Hal was there when Canex became the first trackless mine in Canada. It developed the system by copying the mine at Metaline Falls in Washington State, where management had started using diesel equipment underground. The Canex engineers developed the concept further and eventually became masters at it. Their approach was a combination of that used at Metaline Falls and that used in the Tri-State district in Tennessee. Canex adapted conventional available diesel equipment and evolved it to much larger equipment. This equipment was then used at the Dodger 42 and the Dodger 44.

There were four tungsten mines on the hill (Iron Mountain), and the company shipped their concentrates to Whatshan in the eastern United States. Hal says that the men who made the mining of these deposits practical and profitable were Hub Maxwell, the mine superintendent, and the managing director, Simpson.

The Emerald ore body was a trough like a "V" that plunged at thirty-odd degrees, and Hal designed the shaft used to mine it. The ore deposits were all "contact metamorphic," or very erratic deposits. As Hal explained, one must be almost intuitive to know where the values are. "You don't want to be mining waste," he said, "and Hub Maxwell had this wonderful ability to come up with simple, elegant solutions. He took small pieces of Masonite, a brand of hardboard, and got the shop to bore a specified size of hole through each one. He then gave one piece of this hardboard to each of the shift bosses to carry in their back pockets and said, "If you have any doubt about the quality of the ore, hold this Masonite against the rock and shine your UV lamp on it. If you see six stars there, it's ore." It was simple, and it gave the supervisors control over what they mined."

Phil Graham and Joe Adie. Early 1950s. Courtesy of Marie Adie.

When Hal Kettleson arrived at Canex, zinc was selling for $0.22 per pound. A year later it was selling for $0.08 per pound, but it cost $0.07 per pound to get it to the smelter. It was hard to make money at those prices. At that time the lead paid for the freight and made it possible to keep the mine open. Because of the tungsten contracts, Canex was able to stockpile the zinc and wait for better prices before shipping it. Other mines, which depended on the zinc, couldn't do that, and Remac for example had to shut down. The Canex ore contained about 2% lead and 5% zinc.

During Hal's time at the Emerald there were a total of 122 houses on the mountain, of which about five were at the tungsten concentrator, ten were at the old Emerald mine, and the rest at the Jersey. The Kettlesons lived at the Emerald until they were given a more modern house at the Jersey townsite. Their neighbour there was Bob Hallbauer, who graduated in 1954. Bob and Hal never worked directly with each other, for by then Hal was an underground supervisor, while Bob worked in the office. The Kettlesons shared a duplex with Tony and Nora Triggs, and the Sojas lived in the duplex facing theirs.

George Soja, John Kozar, Ray Jones, and Hal used to go hunting together. They'd each "throw a pack on their backs" and go looking for deer or grouse on the weekends. They also did a little prospecting together.

Hal also remembers Ray Jones, whom he describes as a very personable man with a master's degree in mining and possibly one in metallurgy. He came to Canex as the chief engineer and stayed there for only about three years before moving on to Elliot Lake in Ontario.

Looking down at Cliff's house in foreground of the Jersey townsite. Courtesy of Cliff Rennie.

Nora Triggs, Ann McGowan, Ethel Kipp, and Mildred Stevens with Laura Hill in front.

It seems that at Canex, the senior staff played poker every other Friday evening. Hal was often invited to these games and says he was "not a perpetual loser" but believes that some of the players were invited only because they were. The "regulars" rotated between each other's houses, and the host for the evening usually tried to get one of the non-regulars inebriated "to loosen up his purse strings." Hal remembers a mines inspector who sat in on a poker game. About halfway through the evening he got up, stretched, and said, "Gee, I feel good. I'm up to only $30 down." Thirty dollars was a nice chunk of change in 1953.

Hal Kettleson left the Emerald in the fall of 1956 to mine at Elliot Lake, where his salary was immediately boosted 50% because of his "trackless" experience. His new employer provided him with a new house as well.

When he returned to BC, he became mine superintendent for Ozzie McDonald at Cowichan Copper before moving into construction work with Peter Kiewit Sons Company. A few years later he partnered with a man named Ed Thomas to incorporate "Edco," which specialized in drilling and blasting projects. Hal later sold his share in this company and went into the oil business, where he is still much involved despite being many years past normal retirement age.

RON ERICKSON

Ron grew up beside the Canex lead-zinc concentrator and is the son of the late E.A. Erickson, who took over as mill superintendent there in 1952. It is an understatement to say that Ron Erickson has also been successful, for he and his wife live in a huge, gorgeous house in West Vancouver near Horseshoe Bay. In addition to two apartment buildings in North Vancouver, he also owns another home in Arizona where he goes to escape from some of the North Shore rain. Ron has always been pretty handy with tools, so I was not surprised to learn that he does his own maintenance work on his buildings. Nor was I surprised to see the 1929 Essex car in his garage that he is restoring.

Ron told me that it had been his dream to become a designer and that he actually got as far as applying at the Los Angeles Arts Centre. But a visit to General Motors in Detroit made him change his mind, and he went into business instead. He attributes his positive attitude and his success in life to the way he was brought up—at the Emerald.

Ron and Elsie Erickson, 2007. Ron's father was mill superintendent at the lead-zinc mill for many years.

We lived at the Canex lead-zinc mill, where Dad was the superintendent. It was a good life—an outdoor life. The school didn't have a gymnasium until I was in grade 11. Four of us often cycled from the mill to Salmo during the summer. After school on Friday nights a bunch of us—Gerald Black, Arthur Post, and myself—would grab our packsacks and our rifles, tell Mom, "We'll see you on Sunday night," then head up into the surrounding mountains. We built bivouacs all over that place. We'd cut down some poles and build a lean-to, cover it with fir boughs for a roof, and place more fir boughs below for our beds, then climb into our sleeping bags for the night. We did this not only during the summer but in winter as well. We'd first prepare some firewood for the morning for cooking our breakfast. It was a great way to live. We got lots of fresh air and lots of exercise. It was good clean living—none of us smoked nor drank. It was a simple life. We had no television or telephones. We invented and played our own games.

The senior Evart E. Erickson and his wife Gladys with son Ron celebrate their silver wedding anniversary on February 16, 1961. Courtesy of Ron Erickson.

The Erickson car and their house in the lead-zinc camp. Courtesy of Ron Erickson.

A typical bivouac built by Erickson and friends. Late 1950s. Courtesy of Ron Erickson.

When I got married, I told my wife, who had grown up in Vancouver, "If we have children, I think we should move to a small town where the kids learn how to work with other people, help others, give to and share with others." At the mine we never locked our doors. Friends would come in; if we were having supper, they would sit down, have supper with us. I suggested we do that, but somehow, we never did. Growing up at Canex is a good part of my history that has made me who I am.

When I look at the people I know who grew up in Vancouver, it seems to me that my health is much better than theirs because I've had all that exercise. As a child, we never went to restaurants. My mother cooked every meal at home until I left after grade 12. We usually had a small

Ron Erickson's grade 12 graduating class, late 1950s. Back row, l to r: Peter Pepin, Gerald Black, Ron Erickson, Fred Konkin, Don McDougal, Gaston LeFort, Jack Sheloff. Front row, l to r: Ruby Dodds, Elvina Kraft, Carol Parsons, Mr. John Holden, Althea Treat, Lorraine Stevens. Verna Sauter. Courtesy of Ron Erickson.

Ron Erickson in Vancouver at age 19. Courtesy of Ron Erickson.

garden, so we had fresh vegetables. I look around at the locals here and see the kids eating in MacDonald's or other fast-food restaurants.

BILL AND MOLLY PELIGREN

I first met Bill at Remac in 1955, that is, after his time at the Emerald, when we both served on the Mine Mill Union executive. Bill found out that launching grievances against the company could be risky. Kozar, the mine foreman, had fired a young trammer after a collision between two underground trains. The trammer claimed that he had phoned out first and that Kozar had told him it was okay to proceed. That night there was a dance in the community hall and the fired trammer got into a fight with Kozar. When Bill Peligren tried to intervene, Kozar, at that time still Bill's friend, struck him. Bill laid an assault charge against Kozar, who was found guilty by the local magistrate. Kozar appealed and with the mine's backing won the appeal. He not only fired Bill, but he also had him blacklisted so that he could never work in another mine.

Bill got one more mining job though. It was on the Ripple Rock construction project in Seymour Narrows near Campbell River. Ripple Rock consisted of twin pinnacles that jutted up from the sea floor into the shipping channel between Campbell River and the mainland. Over the years it had been responsible for a large number of ships being damaged or sunk there. The miners drove a tunnel under the ocean and up into the Ripple Rock peaks and honeycombed them with smaller tunnels. When they blasted it with 1400 tons of special Nitramex 2H explosives, it was the largest non-nuclear blast ever detonated.

Bill led off by telling me about how he got into mining.

Bill and Molly Peligren at their golden wedding celebration in March 2002.

I was born on May 5, 1930, at Lumbreck in the Alberta foothills. I left home at 15, and by the age of 17 I was driving truck for Mel Jordy at Keremeos. I had lied about my age so that I could get a truck driver's licence. Mel was terrific and taught me a lot. My first mine was Copper Mountain in 1948 where I worked for about eight months, after which I spent a year at the Giant Yellowknife mine before coming south again. I then worked a couple of years at mines in the Kaslo area, but when I met Molly, I went to work at the Emerald and got a house there as soon as we were married.

Three Boom drill jumbo. Courtesy of Sultan Minerals Inc.

Lost Creek dam—source of water for the mine. Courtesy of Cliff Rennie.

The company used two men to operate the three-boom jumbo. One man was designated as the lead hand. He was responsible, but he was getting the same pay and same bonus as the other man. One day the miners got together and decided that the next time they asked someone to be a lead hand, he should refuse unless he got a higher rate of pay.

The very next time when they needed a man on that jumbo they picked me. I was asked to take over as lead hand because I had a house and three little kiddies, and I guess they thought I couldn't quit. Nevertheless, I refused to unless they paid me more money. I was told by the shift boss to go to the office and pick up my time—I was fired. However, the foreman came and told me I could go to work in the tungsten mine instead. I did that, but a month later, my old friend Kozar from Remac called me and asked me to come down there, so that was when I left the Emerald in the fall of 1955.

Bill's wife Molly, then carried on with her remembrance of events.

I was born in the Nelson hospital but grew up at Thrums. I met Bill and married him on March 31, 1952. We had four children, and the first three were all born while Bill worked at the Emerald. Tim was born while we were still living in Ymir before we got a house at the mine. The house we got at the Emerald was about the same size as the one we later had at Remac, but it had a basement under it. It was one

Above: Fourth Street in Salmo, looking east in the mid-1960s. Courtesy of John Bishop.

Below: The same street looking west in October 2007. Not much has changed in over 40 years.

Jersey townsite. Courtesy of Carol McLean, née Anderson.

Right: Ross and Betty Duthie. Ross went on to become president of Placer Development Ltd. within a few years. Courtesy of Winnie Zuk.

Far right: Charles Arthur Banks, co-founder of Placer Development and Lieutenant-Governor of B.C. from 1946 to 1950. Courtesy of Government House, Victoria, B.C.

Jersey townsite—a different perspective. Photo (looking NW) shot in 1967 from a Cessna aircraft flown by Ed Lawrence. The school can be seen near the top of the picture on the extreme left. The swimming pool can barely be seen very near the top centre of the picture. Courtesy of John Bishop.

Mine Rescue competition. Courtesy of Pat and George Sutherland.

Main camp; general office in rear and engineering office in forefront. Courtesy of Al Nord.

Looking out over the valley and the curling rink below. Courtesy of Winnie Zuk.

The Main Office at the upper (main) camp. Mid-1950s. It had been the camp kitchen with staff quarters upstairs in 1952. The picture below is of the camp commissary. Courtesy of Terry and Barbara Allen.

The tungsten tailings pond in 2007. Note how the trees have reclaimed a good part of it. Courtesy of Sultan Minerals Inc.

Photo of painting that was done by Mrs. Orbeliani (Sr?) during the 1950s. Courtesy of Pamela and Dennis Milburn.

Above: School bus stop (watercolour). Below: The view from Judy Wakefield's window (watercolour).

Photos of Gail Short's paintings done for Judy Wakefield in 1993. They are based on Wakefield photos taken at the mine. Gail is a well-known artist whose work has been shown extensively. She has been teaching Art at Vernon Community Art Centre since 1994.

Looking north from the mine. Highway 3 and a field can be seen at the centre of the photo, and the outskirts of Salmo are visible to the right.

A popular winter sport—snowmobiling. Courtesy of Nancy Verigin.

of those "matchbox" houses. I remember that going to town during the winter was suspenseful—you never knew whether or not you would get back to the mine. I also remember Bill shooting a bear right on our doorstep. We had three children: Tim in 1953, Sandra in 1954, and Ken in 1955. I did not get involved in curling or other athletic events because it not only seemed that I was pregnant the whole time—I literally was.

After we moved into town, when Bill was away, I used to keep the wood furnace going but let it die down during the night. In the morning I would turn on the oven in the electric stove and leave the oven door open to heat the place before getting up and dressed. One time my daughter Sandra came running into the bedroom yelling, "Mommy, Mommy, fire, fire!" I couldn't figure out what she was talking about because I had let the furnace die out. I ran out in time to see that my 2½-year-old son Ken had stuffed the newspaper into the oven where it caught fire from the hot element. From there he dragged it across the room and put it in the toilet. He left a trail of ashes across the floor.

LOUIS PONTI

I knew of Louis long before I met him for this interview. He used to drive the local ambulance in Salmo, and his mother ran a clothing store there. He worked at the mine for two four-year periods: 1952 to 1956 and 1969 to 1973.

Louis Ponti, parts manager at Canex, amongst other things. 2007.

I was born in Blairmore and grew up there and in Salmo. My dad was a miner and a logger. He worked in a coal mine at Blairmore and later at Sheep Creek *[near Salmo]* in both the Queens and the Gold Belt, where he got silicosis. The Gold Belt was particularly bad. The dust was from both the drilling and from dry muck piles. They had water for only a part of the day, and then they had to drill dry. The ironical part of it was that when I worked up there during the Christmas holidays in 1947, I learned that in the mine dry they were blowing in aluminum dust to help these fellows' lungs shed the silica. When you hear about this later on, it just boggles your mind. We now understand that the aluminum dust did more harm than good, but at that time it was thought to

Shop, office, and warehouse. Courtesy of Sultan Minerals Inc.

alleviate the silica problem. I can remember seeing Howard Moore come up and shoot it into the dry during change of shift. They had an air spigot in there and they would put this little can on it, puncture it, and then blow that stuff all through the dry. Needless to say, the whole dry was blacker than the ace of spades—all your clothes as well.

I went to work right after high school in 1948. I was 17 and worked for Nelson Transfer down on Vernon Street. We used to be the GM dealership in Nelson. I believe it later became Reuben Berge Motors. Bill Haldane sold cars there. He was a hockey player, and I used to play ball with him. I left there in 1951 and went back to Blairmore for a year, where I was a parts manager for a GM dealership. That's what I had done at Nelson Transfer.

I came back to Salmo when my mother opened a dress shop here. She expanded it and wanted some help. My wife was pretty good with clothes and she didn't care for the Crows Nest Pass, too windy up there. So we moved back and I worked in renovations at the store quite a bit. Then my dad said, "You'd better come up to Canex." He was working up there on the labour crew. So I went up there in 1952 and worked on the labour crew with old Emil Lund. I had been there only a short time when the company learned I had experience in GM parts—they had a fleet of GM trucks up there, so I got transferred into the transportation department and worked for a fellow named George Walbeck. He was an engineer and he transferred to the engineering office and took me with him, so I worked in the engineering design office for the rest of my stay there.

The underground repair shop at Canex (out of the weather). Courtesy of Pat and George Sutherland.

I did do a little bit of drafting, but I was more of an expediter. There was a lot of construction going on, and they were always looking for shipments of stuff they couldn't find. So that was my job: to get parts. Then they built the underground diesel shop in the Jersey 4200, and one of the draftsmen was put in there in charge of maintenance. I went in there to look after the parts department. We had a pretty fair shop in there.

I never lived at the mine but always stayed in Salmo. I worked at the mine from 1952 to 1956 and again from 1968 or '69 to 1973, just before they closed her down. I had worked in the HB in the locomotive shop from about 1965 to '68 before returning to the Canex. Now they wanted somebody back up at the HB who had experience with the "lokies." So they made arrangements with Ed Lawrence, who was second-in-command then. He said, "Sure, he can go." They had a party for me and I went back to the HB in 1973.

When I went back up to Canex the second time, I was in maintenance. I was in the underground shop for a short time, and then I moved down to the tungsten concentrator when they did the renovations to put it back in service. Then I stayed there as a lead hand supervisor for maintenance in the tungsten concentrator and the underground crushing. They ran that concentrator for about four years, and it was during the time I was there in '52 to '56. But shortly after that the contract with the US government ran out. I think they were getting $60 per unit when the price on the open market was around $28. Naturally the US government did not renew the contract. So they shut the tungsten down. I think the Canex had a crew of about a thousand people then. They were doing a lot of construction. Bunkhouses were being built. They had a little tent city up there on the ball park.

Over the years a lot of people went through there. When I recently talked with Ed Lawrence, he said he wanted to get together with me and we would have a meeting of the locals who had worked up there, and I made up a list of those people and it's not a very long list.

TONY AND NORA TRIGGS

I met the Triggs couple when I interviewed them in 2007 at their North Vancouver home. They provided me with a number of excellent photos for this book. While we were sorting through the photos, Tony said, "I have all the negatives for these pictures, but it would take me a lifetime to sort them out. Unfortunately, I don't have a lifetime left." Tony was an engineer and they lived at the mine from 1952 to 1958.

Nora began by telling me about her early memories of the mine.

Nora and Tony Triggs, B.Sc. (Eng., Mining), now live in North Vancouver. They provided a number of photos for this book. 2007 photo.

Tony, Bob Hallbauer, and Hal Kettleson used to go duck hunting near Creston, so they went off this particular weekend, and while they were away everything froze up at home. Of course when they left the weather was fine. Everything froze! We had only little oil heaters to heat the houses and I was looking after Hallbauers' house because Joan was in Nelson. Her water froze in the sink almost solid because the tap had been left dripping. Then I lost the key to Hallbauers' house just before Tony came home. When he got there, I was looking in the snow bank for the key and I was so mad at him for the fact that they went off hunting and left us holding the bag, looking after the children and our houses in such awful conditions.

I was with Joan [Hallbauer] the time she fell down the stairs at the community hall. I'm the one she accused (in fun) of giving her that push. I thought that she was dead when she hit the bottom, but she had been drinking, so was quite relaxed. Actually, she was about three sheets to the wind.

148 / JEWEL OF THE KOOTENAYS

At the mine we had a great swimming pool for such a small community. It was still there the last time Tony was up there a few years ago. Of course no one is going to take it down. It was a lot of work building it.

I was born in Edmonton. I lived for a few years with my grandparents who farmed near Calmar, Alberta. I came to Nelson when I was three or four. My stepfather's name was Gormley, and he was a miner and a shift boss. He came from a family of about twelve.

We had three children: Patrick and Pamela were born while we were at the mine, and Geoffrey was born in Vancouver after we left the Emerald. Patrick presently lives in Kamloops where he has his own little civil engineering company. Geoffrey and his wife both live in North Vancouver, and Pamela, who married Jaffery Gow, lives in Hunt's Point in Washington State.

Tony then took over with the above hunting experience.

In Creston it got so cold that we could not stay in the tent, but had to go into town and get a motel room. The next morning it was so frigid that we couldn't even get Bob's car started. We pushed it up and down the highway for about an hour before we got it going. Nora always blamed me for the weather on that November 11th weekend, but of course I had no control. The temperature was a balmy 20°F until mid afternoon when it suddenly plunged so quickly that the mud hens on Duck Lake were frozen in the ice. By dark it was below zero with a strong wind blowing, so we abandoned our tents for a motel room in Creston.

Tony Triggs, Vince Killeen, Mill Supt R. McLeod? Hal Kettleson, and Don Wedell. Courtesy of Terry and Alice Sanford.

We returned to the mine the following day at about midnight. Nora was searching for the key. The temperature was −35°, a rare occurrence. The fuel line to our heater was frozen and there was no heat whatsoever in the house.

We had a curling rink and a small skating rink, but we sometimes had to curl only from late night to early morning because although we got a lot of snow, it often wasn't cold enough to maintain the ice in either the skating rink or the curling rink.

I was born in Nelson. My mother came from England to Nelson with her family in 1910. My

Duck hunters: Tony Triggs and Bob Hallbauer. Courtesy of Nora and Tony Triggs.

father came out from England in 1912, and after working his way across Canada at various jobs ended up at Kaslo and the Slocan, where he worked in mines before finally working on the lake boats. When the war ended he returned to the position of purser on the paddle wheelers on Kootenay, Okanagan, and Slocan lakes until the railway was pushed through to Nelson from Kootenay Landing, when he took a position in the ticket office. Nora and I were very attached to Nelson and pleased to be at Canex and so close to it.

I had just finished five years of university, compliments of the government for my three years of service in the Armed Forces. I did not go overseas because I was too young for the first draft—you had to be nineteen, and I was categorized out from two others because of my eye sight.

I met Doug Little at UBC, though he was two years ahead of me. We became good friends at the Emerald, and I sometimes stayed in their house when they were away, while we were waiting for ours to be made ready. Nora had to live with her parents for a bit until we got a house at the mine.

Nora carried on with how she and Tony met.

We met at the beach in Nelson—we all used to go to the park to swim or take along a lunch for a picnic. We had great times there. Then Tony went off to the army and I went to train as a nurse at St. Paul's Hospital in Vancouver. Joan Hallbauer was there too, but a couple of years ahead of me. I say I put Tony through university because we were married a couple of years before he graduated and I nursed for that whole time.

My brother was a photographer in Nelson—entirely different people. He was a trapper and a logger up in the Lardeau area when he got out of high school. Then he sent some pictures he had taken up there down to the Brooks School of Photography, and they offered him a scholarship. He got married, had some children, got divorced, then he went on the tug boats here on the coast and was logging in camps, wrote some songs, made his way across Canada to Montreal, where he got a job as the archivist for the Norman Photographic Files at the McCord Museum.

Tony then talked about his mining career.

I began my mining career at the Hedley Nickel Plate mine when I was seventeen after getting a

Marie Adie, Phil and Phyllis Graham. Courtesy of Nora and Tony Triggs.

letter of introduction from Arthur Lakes in Nelson. It was illegal to work underground below the age of eighteen, but my age somehow gained a year to accommodate this stipulation. I liked the underground work and I didn't like the open pit nearly as much.

At the Emerald I started in the tracked part of the mine where I did all the surveying, measuring, and planning. After about a year I went down to the tungsten mine where I was also surveying until I became a shift boss and then a foreman at the Tungsten in the Dodger 44. It was a trackless operation, and while I was there we sank an inclined shaft at −30° slope. Later on, we were going to sink a shaft just east of the office to the Feeney ore body, but we cancelled that. We eventually accessed the Feeney via a tunnel from underground. When Ray Jones left, I became the chief engineer. During that time I designed and supervised the building of the lead-zinc tailings pond on the flats beside the Salmon River. In 1958 I moved to Craigmont as chief engineer.

Tony Triggs, Vince Killeen, Bob Hallbauer, and Andre Orbeliani in back. Laura Hill and two unidentified people in front row. Courtesy of Nora and Tony Triggs.

There was an electrical superintendent at the mine named Nick Smortchevsky. He was a White Russian who had escaped to Manchuria with his parents from Russia during the revolution. He was educated in China and left it after the war, came to BC to work at the Nickel Plate, and then came to the Emerald. He was a brilliant man, both mechanically and electrically. He lived into his nineties and has a son, Nick, in North Vancouver and a daughter Natalie in Clearwater. *[He also has a son named Adrian.]*

In 1954 a metallurgist named Ross Duthie came over from Riondel to work in the lead-zinc mill. He later transferred to the engineering department to gain mining experience. The first assignment I gave him was to look after a small vertical raise driven beside the lower part of an ore bin that we had to grout to prevent water from entering it. We had a fellow named Harold Steane whom Simpson had brought over from Australia to head up the metallurgical department when we were building the tungsten mill. He together with Jim Eastman and me eventually wound up in the engineering department in Vancouver where we became known as the "Salmo Mafia." There were a lot of people there who were not from Salmo and we were always talking about our days at Salmo, which must have irked some of the others who could never really be a part of that group.

When I arrived in 1952, they were building a conveyor system to replace the aerial tramline. The conveyor came out at the lead-zinc mill

No. 2 conveyor—a different perspective. Note that the belt is covered with galvanized iron. Courtesy of Al Nord.

at the bottom. Then over the years a water problem developed and there was drainage into the raises which caused all the blowouts. So we had a big grouting program on. It was still a lot better than the tramline, because on it the buckets would sometimes run away and crash into each other and scatter muck all over the hillside. The steel towers that replaced the wooden ones were sold to build the Salmo ski tow later on. The mill below was originally built for tungsten and then converted to lead-zinc with much higher capacity. Later they built a new tungsten mill up above. We were still building the conveyor system when the managing director, Mr. Simpson, came up. He asked, "What is all that aluminum cladding on the sides of the conveyor way? We are a zinc property and we should be using zinc."

When I went up there in 1952, they were in transition. They had some guys in from Princeton who were running the property. I think Harold Lakes was officially the general manager, but he wasn't well. Things were bad, poor discipline. The tunnel crew had Eimco overhead muckers, and they'd be out there throwing rocks to each other with them. That's when Gerry Gordon was the manager of the tracked mine. He had come down from the Pioneer mine, and so they had a sweep of those guys, got them out, and Gerry became the general manager of the property, and they got rid of all those Princeton guys. Gerry was a good manager. Doug Little became superintendent of the tracked mine at that time. A fellow named McCutcheon became superintendent of the trackless. I believe that Jack Anderberg was one

Gerry Gordon, Terry Sanford, Bob Weber, Phil Graham, and Harold Wainwright. Courtesy of Nora and Tony Triggs.

of the fellows that left during the house cleaning. It had gotten out of hand because there was so much going on. We were building the tungsten mill and we were doing the conveyor system as well as an underground crusher. Furthermore, that trackless mining was a new thing.

Ross Duthie and I moved on to Craigmont at about the same time and later to the Vancouver office, where he eventually became president of Placer. Then we worked in the Philippines—I took the family there for a few months one summer, but we came back three years later. We were supposed to be there for only five months, but Doug would come down and say, "Will you stay one more year?" And then it was another year, and so on. We always felt rather unsettled because if you go for three years, it's one thing, but going for a few months at a time is very different. The people were lovely, so generous, always giving. We'd try to give them something back, but then they would come back to us yet again. After we came back to Canada, I worked out of the Vancouver office, building mines in New Guinea and Australia, whatnot. Recently they sold our entire company to Barrick Gold.

TONY GIZA

I first met Tony in 1955 when I worked at the HB Mine. He was a big, strong, husky miner who worked at Canex from 1952 to 1960, with some other jobs in between. I encountered him again at Ymir during the union picnic in 1956—I was now at Remac and married.

One event at the picnic was shovelling one ton of broken rock from one compartment into another in a long wooden box built for that purpose. Big Nick Bolinsky from Remac was a shoo-in to win the competition, but I thought I might be good enough to take second prize—there was no third. I made a strategic blunder by being one of the first competitors. I worked so

Tony Giza of Salmo, 2006.

hard that I was ill for the rest of the day. By the time Tony took his turn, a fair amount of the muck had been spilled outside the box, and Tony beat me for second place by a mere three seconds. Tony is the kind of guy you like for a partner. He is quiet and unassuming, but a top notch worker and friend. Here is what Tony told me:

I was born in 1924 at Blairmore—a town of about 3,000, where my dad worked in the coal mine. The neighbouring towns of Coleman, Michel, etc. all had mines as well, with Coleman having three.

My dad had eight children and later I had six of my own—five girls and a boy. They have in turn produced twelve grandchildren and four great-grandchildren up to now. My wife passed away in 1992. My children all returned for the Salmo Homecoming reunion in 2006.

My work underground began at age 18 in 1942, but I joined the army the following year, but never made it to Europe. After getting out of the army in 1945, I married and worked in the coal mines again until I came to Salmo in 1949. Within four days of saying I would never work in another mine, I hired on at Remac and worked there for about 14 months. I lived in Ymir and worked at Canex most of the time from 1952 to 1960.

At Canex I worked in the Dodger 42 tunnel while Bill Peters was the shifter. It was Peters himself who drove the Dart to Salmo for beer. In the tunnel we even had a sump full of water where they kept the beer cool. It was goddam lucky they didn't hang us all. We had a hell of a good time that day. Lloyd Johnson was on our crew and was practising on an Eimco overhead loader. He came out the portal and dug a great big hole and tipped the sonovabitch right over in that hole. He wasn't even the operator, but was trying to learn to run it. We had Mike Posnikoff, Jimmy Ash, Bill Peters, Dunc McDonald, Lloyd Johnson, and I on our crew. Dunc normally ran the loader.

On another occasion, two of the guys were playing catch with two Eimco overhead loaders—throwing a big rock back and forth between them. Jack Anderberg was still the foreman at the Dodger then, and it was probably because of antics such as these that management cleaned house and got rid of him and some of the other guys.

Someone had stolen a stoper [drill] at the Emerald and had it in the Ymir beer parlour one evening. Meanwhile someone else had stolen a stoper at a different mine and had sold it to Old Tarkman and Patulla at the Good Enough mine. The police who were looking for the Emerald drill, got wind of it and went looking for it there. They got the stoper all right, but it was the wrong one. *[There were close to a thousand men at Canex at the time and theft was rampant. Whole spools of wire rope would go missing, not to mention the number of small tools that regularly went home in lunch buckets. Most miners had pipe wrenches, hatchets and crescent wrenches they borrowed at work.]*

This is truck is similar to the one Bill Peters drove to Salmo for a load of beer. This is the adit in which I worked during the summer of 1952.

I went to work outside in the spring one year—I wanted to get out of the mine for a while. They put me to work on this big silo that was going up. I knew the bosses pretty good, eh? I stayed there about three weeks. We were up on that goddam silo putting the roof on it. Holy Fuck! That was in March and the sun came out and I couldn't hack her. I was up on that goddam roof and walking around on that little scaffold. It was hot and the height bothered the hell out of me too. We were a long way from the ground—probably a hundred feet. And yet in the mine and I'm hanging on a goddam piece of rotten rope hundreds of feet up in a raise and it don't bother me at all. I told old Gunnar Adolphson, "You gotta put me back in the mine." He said, "I knew damn well you were not going to stay out here too long." He was looking after the *[Dodger]* 42 then as well the Jersey. That was quite a goddam mine to work in. A lot of people worked there for a long time. I stayed there until 1960 when the PF Law Company put in the first three miles of the new Creston highway south of town.

The following year I worked for Emil Anderson on the section 10 miles up to the summit. The highway was open in July of '62. I know because I drove over it to go to Sparwood. There was a lot of rock on that job.

Although I was at the Emerald from '52 to '60, it was not in one single stretch. I don't know how many times I quit the Emerald, but I always went back. One time I went to work in a gopher hole at Ainsworth—the Scranton mine. It was up at the top of Woodbury Creek. Bob Golac was another one I worked for several times. He was good to work for—too good. He also did some tunnels for the conveyor belts at Canex.

I also worked for Paddy Harrison in a couple of tunnels over at Michel. They had a coal seam that caught fire, so we drove a tunnel to help get the fire under control. We drove a decline as well and I was on the Scooptram. It was so steep (29%) that even with the tires chained up I would spin out coming up the ramp with a load. It was so bad that we had to use a slusher hoist to help pull the Scooptram up the grade. That made it a pretty slow operation.

I'll never forget my first day underground. Holy Christ! It was in the coal mine at Blairmore and they told me to go with this guy packing timber. You had to pack timber for the miners for six months before they would let you work with a miner. After that you could work as a helper with a first-class miner to learn how to mine. We had to go up a raise, along a drift, through a crosscut and up another raise to surface where we threw timber down the raise. I had no idea of where I was, it was so goddam black in there and that little light didn't let you see very far. We would throw a bunch of timber down this raise, then from there we would carry it to the various miners who would each go through about eight timbers a day.

Unknown, Gunnar Adolphson, and Phil Graham. Courtesy of Nora and Tony Triggs.

Anyhow, we get down there and this guy I'm working with gets into an argument with the fire boss and he quits and leaves me there. The fire boss said, "You keep on workin and get some timber down." Well, I'll tell ya, I got turned around in that son-of-a-bitchin' mine and blacker than a bastard, and I don't know where the hell to go. So I said, the hell with it. As soon as he buggered off, I buggered off too. I finally got down to the tunnel where the tracks were, but I didn't know which way to go. That scared the piss right outta me. Anyhow, I followed the tracks and they took me outside, but that tunnel went from one side of the mountain to the other and I went out the wrong side. That tunnel was about 2½ miles long and it took me almost all day to get out and I never run into anybody. I'll tell you, I never left my goddam partner after that. If he went someplace I would be right with him.

Hazing was a normal part of being a greenhorn in the mine. Some big miners would hold the new guy down and smear his balls with track grease, or they might instead send a guy on a fool's errand. Everyone got initiated one way or another. There wasn't a goddamn thing you could do about it.

I asked Tony, "If you had to do it all over again, would you do it any differently?" He replied, "Well, I think I would stay out of the mines if I had a chance." However, he conceded that the mines had been good to him. He now lives alone in a big house but says he doesn't talk to himself for two reasons: there is no one to answer him, and without his hearing aids he couldn't hear it anyhow.

TERRY AND ALICE SANFORD

Terry Sanford was an electrician at the Emerald from 1952 to 1959. I met him and his wife in their home at Robson. In addition to his electrical skills, Terry must have been a good renovator as well, for although the house was a shack when they moved into it, today there is no sign of it ever having been anything but a lovely modern home. The Sanfords showed me quite a number of photos and then Terry started to talk.

I was born in 1920 near Windsor, Nova Scotia, and during the war I spent five years in the Armed Forces, where I served in medium-sized gunnery. After the war I worked at the Giant Mine near Yellowknife, where I apprenticed as an electrician.

In 1952 I moved to the Emerald, where we stayed until the strike in 1959. After one day on the picket line I decided it was not for me and found a job at the Columbia River dam near Castlegar. Alice stayed in the company house for another three weeks before she joined me there. That was kind of a dicey situation, with me no longer being on the Canex payroll. Meanwhile, I had bought a shack at nearby Robson, in which we lived as we gradually improved it over the years. We have now been in that house for 48 years.

Terry and Alice Sanford, 2007. Terry was an electrician at the mine until strike duty led him to try construction work instead.

Alice then chipped in:

I was born at Viking, Alberta, and worked in Yellowknife as a nurse and met Terry there. We were married in 1952 and left the Territories for Terry to work at the Emerald mine near Salmo. There we moved into a two-bedroom duplex. It was $25 per month and it was cozy, but I remember that when visiting some friends in a Panabode, we kept our coats on.

When the first TV sets came to the mine, we used to visit with someone who had one and watch "Gunsmoke." We had three children, two daughters and a son, born in either Nelson or Trail hospitals while we were at the mine. A son

Peggy Rowe, Alice Sanford, Mildred Stevens, and Rhea Colwell. Courtesy of Terry and Alice Sanford.

Proud curlers: Mary Copley, Alice Sanford, Rhea Colwell, and unknown. Courtesy of Terry and Alice Sanford.

who is now 47 was born after we moved to Robson. He is a millwright in Kitimat and was just recently married.

What I mostly remember about the Emerald was the deep snow and that it was a wonderful place to live. There was no dividing line between staff and hourly paid people. No one had any money, but everyone got along so well. One sport we loved up there was curling, in which we once won a trophy.

WINNIE ZUK

Winnie is the widow of Steve Zuk, who was a mechanical foreman at Canex from 1952 until 1960. She presently lives in Kelowna.

I grew up in Mission, graduated from high school, and then went to Vancouver for another year of study in fabrics and design. There I met the twin sister of my future husband Steve. She sent a photo of us to him overseas. He wrote back and asked if the girl in the picture would write to him—which I did. I met him in September 1945 when he came to Vancouver for his discharge from the Navy. We were married a year later.

Steve was raised in Trail and when he returned from overseas he was offered his job back with Cominco. When I first saw Trail, I was shocked that any place could be so barren and ugly. I had come from the greenery of the coast, and the vegetation around Trail was dead from the effects of the smelter smoke.

When he asked for a transfer, Steve was given a choice of going to either the Bluebell mine at Riondel or to the Con mine at Yellowknife, so he chose the Bluebell. I was left in Trail as there was no accommodation for wives at the Bluebell. He had quite a trip home on weekends. It involved finding a boat to take him across Kootenay Lake, and catching a bus to Trail—not a very satisfactory situation.

In 1948 Steve applied for a job with the Island Mountain Mine in Wells, BC, and went up to look things over. We had to check a map to find where it was.

I joined him a month later. The road was

Steve in profile. Swimming pool below. Courtesy of Winnie Zuk.

gravelled from Lac la Hache to Quesnel, from where I had another two-hour trip in the "coach" to Wells.

This was the first time I experienced COLD weather. In the winter of 1948/49 the thermometer dropped to −58°F [−50°C]. However, we were in a cozy little log cabin. Gerry Gordon was manager of the Gold Quartz Mine in Wells at that time. We later met Gwen and Gerry again at Emerald Mine near Salmo. We survived the north until 1952, when we returned to the Kootenays and Steve started work at Emerald Mine.

At Canex, we lived in a company duplex near the highway at the lead-zinc mill for the first six months. Then Steve was put on staff, and we were able to move into a company Panabode up at the mine site. I loved that little house even though it was like an oven the summer and our clothes froze to the walls of the closet in winter.

Zuk's house in winter. Courtesy of Winnie Zuk.

Steve wanted to improve the place right away. He used scrap lumber from the mine to build a picket fence, a gate, a sidewalk, and a carport. He next brought up loads of sod over the 22 switchbacks, from a field in Salmo and made a lawn for our yard.

We had a library in the community hall, a curling rink, a two-room primary school, and later on our wonderful swimming pool. Most of us bought our first TVs while living there, for we could pick up KLXY put of Spokane—on "rabbit ears" of course. No one went out on the nights of "Playhouse 90" and "The Ed Sullivan Show." On summer evenings we sometimes drove to the dump and watched the antics of the bears for entertainment.

We usually went down the hill to Salmo twice a month for groceries and to the bank. Dr. Carpenter was there and also the drugstore. In the summer a Chinese man known as Mr. Bing would come up the hill weekly with fresh fruit and vegetables. We learned to be very resourceful and exchanged many wonderful recipes, some of which I still use today. We all made our own bread. Steve would go huckleberry picking near Nelson in the summer. These would be preserved or made into jam. There is nothing to compare

Two Zuk children learning hockey. Courtesy of Winnie Zuk.

JEWEL OF THE KOOTENAYS / 159

with a huckleberry pie! In late summer we would bring peaches, pears, plums, and tomatoes back from the Okanagan and I would be at the sink for a week trying to preserve all this bounty.

There was a small building near the school known as the "commissary," where we picked up our mail and could buy an emergency can of soup or milk.

Our son Michael was about two years old when we moved to Emerald Mine. Our daughter Nancy was born in 1956—part of the population explosion at the mine. There were so many boys—Betty Rennie with six and Betty Duthie with three; so we were hoping for a girl. There was a lot of pressure on me from my friends to produce a girl, which I'm happy to say I did. The children had many friends to play with and a very safe area to explore. However, we had to keep watch for bears coming near our houses.

Polio vaccine became available for the first time while we were at the mine. We had to take our children to the school for their shots. I couldn't bear to see my son get his needle, so Betty Duthie kindly offered to take him up when his turn came.

We had wonderful neighbours on our street. There were the Ashes, Claytons, Andre and Irene Orbeliani, Stan and Laura Hill, the Martins, Hogarths, Tom and Enid Smith, Ev and Mel Olsen, Norm and Doreen Steele, Ross and Betty Duthie, the Allens, and the Peters.

Many of us were moved over to Placer's new open pit mine in Merritt in 1959/60. This was great, as we had our friends for support while learning to fit into a new area. We were now able to go shopping in local stores—what a treat! My friend Betty Duthie was once again a neighbour and Doreen Steele was just down the road from me in Lower Nicola.

In 1968 we made still another move. We left Craigmont Mine to join Brenda Mine in Peachland. In January 1982 Steve suffered a heart attack and was off work for several months. He took early retirement in September 1983. We moved into Kelowna in 1987 and settled into our new home. During the next few years we travelled extensively. Steve loved to cruise, and we took several wonderful trips that way.

We planned a long trip for our 50th Wedding Anniversary in May 1996. We flew to London, spent time with friends in Britain, and then boarded a ship in Dover. We visited Germany, Denmark, Finland, Norway, Sweden, and St.

Underground shop entrance. Courtesy of Winnie Zuk.

Petersburg, Russia. We had a wonderful time and flew home from London in mid June. Two days after our return, Steve passed away from a massive heart attack.

We had many happy years together and made lifelong friends at Canex. Doreen Steele and I travelled to Britain in 1973, and we kept in touch until her death. Betty Duthie and I still correspond periodically, and the Orbelianis used to visit us in Peachland.

I think we learned many of life's lessons at Canex, such as the value of good friends and neighbours, how to make do with what we had, and how to improvise when necessary. I wouldn't change the years we spent at Emerald Mine for anything.

Winnie and Steve with their grandchildren. Courtesy of Winnie Zuk.

LAWRENCE BOND

What I remember about Lawrence is that he was always involved in union activities. I did not know him well, but he never missed joint meetings between the sub-locals of the union. He worked as a miner at the Emerald from 1952 to 1973.

I was born in Nelson on January 12, 1928, and lived in Ymir until I was about thirteen. We moved to the coast for about five years during the war and lived in both the Vancouver and New Westminster areas.

When I was old enough, I went to work for Swift Canadian on the killing floor for a couple of years. I didn't cut much meat, just worked in slaughtering, skinning, and cleaning. We killed cattle with an eight-pound sledge hammer in those days. Today you are required to use a stun gun. When I quit they said I could come back any time. It costs money to train someone in that business; every cut in the hide you make by mistake costs money. They used everything in the animal. I suspect that they even recorded the pigs' squeals for sale to the movies.

When I returned to the Kootenays I first worked around Retallack for a couple of summers at the Court Province and the Silversmith mines. I also worked outside at the Bluebell for a couple of months before Cominco reopened the mine. It was a typical Cominco operation. If you worked for

Lawrence Bond.

them long enough, they had you so brainwashed that you had only one line of thinking.

I came to the Emerald in October 1952, where I worked briefly in the Dodger before they transferred me to the lead-zinc. When my back eventually gave out on me, I went to work in the mill, but I spent seventeen years as a miner first. I lived in the Jersey townsite for ten years, from 1952 to 1962, and four of my six children were born while we were there.

I worked in a one-man drift, a five-by-seven-foot heading. After finishing the drift, I would slash it out and then bench it down. Sometimes I would have a hundred 20-foot holes drilled before I blasted. In the drift I scraped the muck out using a slusher hoist. Occasionally the cables would break. Like most miners at the time, I fixed broken cables with Molly-Hogan splices. We cut the cables using the traditional miner's cable cutting tools—an axe and a hammer. We set the axe down on solid rock with the sharp side up, laid the cable across the blade, and hit it with a single jack hammer. We did have proper cable cutters, but few used them because you had to run to the shop to get one, and the axe was always handy. It was the miner's favourite tool. I used to take my axe into the shop and get it sharpened more like a chisel than with the normal long taper that was standard for the blade.

I did have one close call. Paul Koochin and I were drilling in an open pit, off the back of an old Dodge power wagon. We took turns drilling, and while Paul was on the drill, I happened to notice that the pit wall seemed to be coming towards us. I grabbed Paul and yelled, "Get out, get out!" He jumped, but his foot hooked in something, so I stopped and grabbed him and got his foot loose. By then it was too late for me to jump, so I just flopped myself over the cab of the truck. When everything died down there was rock strewn all around us. Somehow, neither one of us had been hurt. That was about the only close call I can remember having, but I was always pretty careful. Yet, if we had had those camcorders that are now available, we could have made some of the best movies around. Of course we would have had to censor out some of the language used.

We had a few comical characters at the mine. One guy was always

Sequence of photos showing a miner's typical cable splice. His tools were an axe and a hammer. 2007.

telling us how he was the boss at home. One day when he came into the bar to cash a cheque, he stopped and had a beer with us as well. His wife who had been waiting for him in the car, came storming in and dragged him out. As he was leaving, the guys all yelled, "Now we know who the boss really is."

Big Chris Christensen was about the strongest man I ever saw, though he wasn't that tall, maybe five-ten. In the chutes where the trucks loaded we had these huge chains designed to control the muck flow. Chris and I had to hang some chains from eye bolts above the chute. He said to me, "Here, hold this chain." I'm lying on my back in there with little room to manoeuvre and the chain was just too heavy for me. Chris says, "You wouldn't even make a good babysitter." I replied, "You big bugger, we aren't all as strong as you." Back in the days when we drilled with leyners, he wouldn't even unhook the hoses, but would carry the bar with arm and leyner still attached, and the hoses dragging, to the next set-up. The whole works must have weighed 400 pounds. He could lift a barrel of gasoline onto the back of a pickup truck. The gasoline alone would have weighed 360 pounds, not including the weight of the barrel. Then

APPENDIX "A" — WAGE RATES INCLUDING SAVINGS PLAN
Effective 1 December 1965

Classification		Classification		Classification	
Jryman. Blacksmith	$2.87	Rprmn 2nd - Carpenter	$2.55½	Trackman	$2.48
Jryman. Carpenter	2.87	Rprmn 2nd - Electrician	2.55½	Pipeman	2.48
Jrym. Electrician	2.87	Rprmn 2nd - Machinist	2.55½	Painter - 2nd	2.48
Jrym. Machinist	2.87	Rprmn 2nd - Mechanic	2.55½	Compressor Operator	2.38½
Jrym. Mechanic	2.87	Rprmn 2nd - Welder	2.55½	Crusherman - Mill	2.38½
Jrym. Welder	2.87	Skiptender	2.55½	Rprmn 4th - Blacksmith	2.38½
Bulldozer Operator	2.69	Timberman	2.55½	Rprmn 4th - Carpenter	2.38½
Hoistman - Main	2.69	Truck Driver - U.G.	2.55½	Rprmn 4th - Electrician	2.38½
Painter - 1st	2.76	Ball Mill Operator	2.48	Rprmn 4th - Machinist	2.38½
Rprmn 1st - Blacksmith	2.76	Bucker	2.48	Rprmn 4th - Mechanic	2.38½
Rprmn 1st - Carpenter	2.76	Canning Operator	2.48	Rprmn 4th - Welder	2.38½
Rprmn 1st - Electrician	2.76	Table and Iron Flotation Op.	2.48	Truck Driver - Surface	2.38½
Rprmn 1st - Machinist	2.76	Motorman	2.48	Conveyorman	2.29
Rprmn 1st - Mechanic	2.76	Mucking Machine Operator	2.48	Miner's Helper	2.29
Rprmn 1st - Welder	2.76	Powderman	2.48	Nipper	2.29
Special Timberman	2.69	Rprmn 3rd - Blacksmith	2.48	Pipeman's Helper	2.29
Crusherman - Underground	2.55½	Rprmn 3rd - Carpenter	2.48	Timberman's Helper	2.29
Flotation Operator L-Z	2.55½	Rprmn 3rd - Electrician	2.48	Sampler	2.29
Flotation Operator WO3	2.55½	Rprmn 3rd - Machinist	2.48	Labourer - U.G.	2.22
Hoistman - Sinking	2.55½	Rprmn 3rd - Mechanic	2.48	Mill Helper	2.22
Leach Operator - WO3	2.55½	Rprmn 3rd - Welder	2.48	Tailings Pond Operator	2.22
Loader Operator - Diesel	2.55½	Slusherman	2.48	Tradesman's Helper	2.22
Miner	2.55½	Steam Engineer	2.48	Dry Man	2.16
Rprmn 2nd - Blacksmith	2.55½	Steel Sharpener - U.G.	2.48	Labourer	2.16

Lead Hands to be paid a minimum of six cents (6c) per hour above their normal classification rate. The appointment of Lead Hands is solely at the discretion of the Company Management.

NOTE: Categories 1 and 2 include tradesmen adjustment.
All categories include 5c Company contribution to savings plan, however it is understood that employees who do not qualify for the Company contribution to the savings plan as designated in Article 61 will be paid rates 5c per hour less than the amounts shown.

APPENDIX "B" — WAGE RATES INCLUDING SAVINGS PLAN
Effective 1 December 1966

Classification		Classification		Classification	
Jrymn. Blacksmith	$3.04	Rprmn. 2nd - Carpenter	$2.70½	Trackman	$2.62½
Jrymn. Carpenter	3.04	Rprmn 2nd - Electrician	2.70½	Pipeman	2.62½
Jrymn. Electrician	3.04	Rprmn. 2nd - Machinist	2.70½	Painter - 2nd	2.62½
Jrymn. Machinist	3.04	Rprmn. 2nd - Mechanic	2.70½	Compressor Operator	2.52½
Jrymn. Mechanic	3.04	Rprmn. 2nd - Welder	2.70½	Crusherman - Mill	2.52½
Jrymn. Welder	3.04	Skiptender	2.70½	Rprmn. 4th - Blacksmith	2.52½
Bulldozer Operator	2.85	Timberman	2.70½	Rprmn. 4th - Carpenter	2.52½
Hoistman - Main	2.85	Truck Driver - U.G.	2.70½	Rprmn. 4th - Electrician	2.52½
Painter - 1st	2.92½	Ball Mill Operator	2.62½	Rprmn. 4th - Machinist	2.52½
Rprmn. 1st - Blacksmith	2.92½	Bucker	2.62½	Rprmn. 4th - Mechanic	2.52½
Rprmn. 1st - Carpenter	2.92½	Canning Operator	2.62½	Rprmn. 4th - Welder	2.52½
Rprmn. 1st - Electrician	2.92½	Table and Iron Flotation Oper.	2.62½	Truck Driver - Surface	2.52½
Rprmn. 1st - Machinist	2.92½	Motorman	2.62½	Conveyorman	2.42½
Rprmn. 1st - Mechanic	2.92½	Mucking Machine Operator	2.62½	Miner's Helper	2.42½
Rprmn. 1st - Welder	2.92½	Powderman	2.62½	Nipper	2.42½
Special Timberman	2.85	Rprmn. 3rd - Blacksmith	2.62½	Pipeman's Helper	2.42½
Crusherman - Underground	2.70½	Rprmn. 3rd - Carpenter	2.62½	Timberman's Helper	2.42½
Flotation Operator L-Z	2.70½	Rprmn. 3rd - Electrician	2.62½	Sampler	2.42½
Flotation Operator WO3	2.70½	Rprmn. 3rd - Machinist	2.62½	Labourer - U.G.	2.35
Hoistman - Sinking	2.70½	Rprmn. 3rd - Mechanic	2.62½	Mill Helper	2.35
Leach Operator - WO3	2.70½	Rprmn. 3rd - Welder	2.62½	Tailings Pond Operator	2.35
Loader Operator - Diesel	2.70½	Slusherman	2.62½	Tradesman's Helper	2.35
Miner	2.70½	Steam Engineer	2.62½	Dry Man	2.28½
Rprmn. 2nd - Blacksmith	2.70½	Steel Sharpener - U.G.	2.62½	Labourer	2.28½

Lead Hands to be paid a minimum of six cents (6c) per hour above their normal classification rate. The appointment of Lead Hands is solely at the discretion of the Company Management.

NOTE: Categories 1 and 2 include tradesmen adjustment.
All categories include 5c Company contribution to savings plan, however it is understood that employees who do not qualify for the Company contribution to the savings plan as designated in Article 61 will be paid rates 5c per hour less than the amounts shown.

Mine Mill pay scales. Courtesy of Andy Wingerak.

you wonder why so many of these guys wind up with bad backs later on. But what good is it to be as strong as a horse if you can't show off and impress people?

We had a guy once who got a contract price based on how many feet he drilled. He went into the stope and drilled holes a foot apart and got very good footage. When I was sent in to blast them, I didn't even load one-half of them. I asked him afterwards, "How come you drilled so many holes?" He replied in his heavy accent, "I drreel holes, I get paid!"

Harvey Murphy of the Mine Mill Union was an avowed Communist, but I never ever heard him try to influence any union business politically, and I spent many hours in meetings with him. Around here guys would say, "It's no use running for the union executive. It's Forgaard, Walton, and Bond running the show." I used to tell them, "Well, why don't you come to the meetings?" We had a combined membership of over 400 between the three mines in the area, and at the meetings, possibly ten people would show up.

We had a meeting up here one time and we were drawing up proposals for the bargaining committee. We were going over the things we thought we should be asking for, when someone piped up, "What about overtime?" I said, "What do you mean?" He says, "Well, how much do we get for overtime, time-and-a-half, two times, two-and-a half times?" I said, "No, one-half the base rate." He says, "Well, then, no one will work overtime." I said, "That's the idea. If they need you to work overtime, maybe they should hire more people instead." That was the union's philosophy. He didn't think much of that. After the business part of the meeting was over, I would say, "Does anyone here have any complaints, because if you do, I want to hear them now, not tomorrow or in the beer parlour." Once in a while someone would get up and we would have an argument, but not too often.

Our business agent in Salmo, Bill Rudichuk, was the best business agent that ever was. He went out of his way to help people all the time. I know a couple of old-timers here who couldn't get their silicosis pensions. Rudichuk told them, "Bring in all of your information." He got them pensions or raises in their pensions and stuff like that.

I remember the time in the tungsten; they were drilling and hit a vein of high-grade silver. It was only a sliver I guess. Well, they went crazy. They were drilling everywhere. The silver in the area was usually in pockets, not true veins, but it was usually rich. I know there was one place up there where there was a big kidney deposit on the side of the hill and they were rawhiding it with horses, but that's all there was, just that one pocket.

When Mine Mill merged with the Steelworkers Union, it was a fixed deal. The Canadian Labour Congress had sold Mine Mill's jurisdiction to the Steelworkers for $50,000. A lot of people didn't know that.

I spent some years in the tungsten mill. I knew a little about milling before I went but learned most of it after I came. Tungsten milling was different from lead-zinc. In the tungsten you separated most of it, but then you had to roast it to drive the impurities out of it, otherwise the buyers wouldn't touch it. We used acid, oh Christ, the fumes from that acid.... The roaster would sometimes go haywire and you'd get the sulphur dioxide gas. You got one whiff of that and your lungs would just burn. I used to hold my breath when I had to run through it.

I remember one time they shut the roaster down, and then when the mine inspector showed up, I said to the inspector, "Oh, now I know why the roaster shut down. You let the company know you were coming, didn't you?" When he protested, I said, "You're damn right you did. How would they know to shut it down if you hadn't?"

I recently got new hearing aids, courtesy of the WCB. They are about my third or fourth set. While I mined there were no such things as ear plugs, and in the raises, those stopers used to have pretty loud barks.

THE STENZEL FAMILY

The Stenzel family in front of their house in 1953. Courtesy of Ernie Stenzel.

I met the Stenzel family at their May 2007 reunion at Elaine Gorsline's home in Vernon and was able to interview Elaine and her younger brother, Ernie.

Elaine's father, Frank Stenzel, had died suddenly, leaving his widow, Mary, with several children and no job. To complicate matters, they were living in a company house. Mel Olson, the chief accountant, met with senior management and suggested if Mary were to go to school and take a secretarial course, he would hire her. They charged for neither rent nor utilities during that period of time. When I interviewed Mel Olson, he said that Mary Stenzel was the best secretary he ever had.

ELAINE GORSLINE, NÉE STENZEL

We lived in Ymir and Dad stayed in the bunkhouse and came home on weekends until we moved up to the mine in 1952. I was eight years old and lived there for the next ten years.

When I was a young girl growing up there, I was never accosted by any of the men living in the bunkhouses. I don't remember ever being afraid. Just the same, delivering newspapers to the bunkhouse was one job no girl ever got. Something I will never forget is the smell of Lifebuoy soap. Whenever I walked by the bunkhouse I could smell that soap.

People moved around a lot up at the mine. It seemed to me that someone was always moving from one house to another. I guess if a house came open, someone would try to move into it if they thought it was a better one.

Mel Olson was my mother's boss while she worked in the office. She had nothing but good words for him. I think he could be tough and very particular, but Mom enjoyed that because she liked her work to be meticulous, and she really appreciated all the support he and the management extended to her when Dad died. She enjoyed working for him even though some others didn't. She used to drive to Trail every day to take her courses.

When we first moved up to the Emerald mine, we lived at Main Camp. There were about a dozen families living there. That was when television first came in and Elaine and Colonel Perry were one of the few households on the hill that had TV. They used to invite all the children of the camp to come and watch "Lassie" and G.E. theatre at the Perry house every Sunday. The entire living-room floor would be covered with children, who all had to behave themselves. It was wonderful! The Perrys were wonderful people to invite all these kids.

Elaine Gorsline, née Stenzel, of Vernon was hosting a family reunion when this picture was taken in 2007.

I remember being driven to the Harold Lakes School in a panel truck that was also used to deliver groceries, mail, and whatever. The vehicle had plank seats down each side of it. We kids would pile in and sit on those planks and slide back and forth on them. And it was dusty! I wonder what the school board would think today about hauling kids around in such a vehicle. That panel truck did good things. Once when it drove by, I saw a toboggan on its roof that was being delivered to a neighbour. However, it was under our Christmas tree the next morning. Mom must have ordered it from Sears. That panel truck brought the mail, it brought prescriptions, and it brought toys. It left me with a warm memory of it.

Frank Stenzel, Ernie's father. Courtesy of Ernie Stenzel.

When I went to grade 7 in Salmo, Mr. Phelps drove the school bus. One morning during our ride on the bus, a front wheel came off and went rolling down the hill ahead of us. I'm sure a few people will remember it. Mr. Phelps just turned around and said to us, "Everybody stay on the bus." He got off the bus, hitchhiked to Salmo, and came back with another bus and got us to school on time.

One memory I have is of the blasting whistle *[there was an "open pit" mine above the main camp]*. We would sometimes have company when the whistle would go off. Our guest would exclaim, "What's that?"

We'd reply nonchalantly, "Oh, that's just the blasting whistle." Then the shot would go off with a loud whomp and we would see rocks flying through the air and landing in the trees. We loved it because we knew what was coming and our visitors did not. They would invariably exclaim, "Oh, what was that?"

The company was pretty good. They had turned the old cookhouse down at Tungsten into a community hall. We celebrated Halloween there, had teen dances, and in general had great times there.

One time I was down at the Salmo high school and had missed the school bus. My parents didn't have a car and I was wondering, "How am I going to get home?" I telephoned Mrs. Gordon, the mine manager's wife and she drove down and picked me up and took me home.

I started school at Canex at the end of grade 4, so I was in school for only a month. We had a teacher, Mrs. Murray, who at the end of the year gave each student a little present, but she did not have one for me because I arrived so late in the term. I'll be darned, but sometime in July I got a letter from Mrs. Murray and a little handkerchief. She realized that I had not gotten a gift and I guess she didn't want me to feel left out.

Every family got paid on the same day, and we would all go into Nelson to buy groceries and on the following day, in winter, the discussion would be about who had made it up the hill, who had to put the chains on, who had to get pulled out of the snow bank, and who did not make it back at all. If we were in Salmo or Ymir, and you said, "I'm going up the hill," everyone understood you to mean the Emerald mine.

My dad liked to do little things for people. One time when someone had a child with a cast on his hip, Dad made a little carrying chair for it. On another time when the grandmother of Smortchevsky passed away, they asked Dad to make a Greek Orthodox wooden cross for them. Little four-year-old Ricky Fritz came over and said, "Mister, are you making a cross for the angels?"

When we had the Emerald reunion, we did not even think to invite Dr. Carpenter, but he came anyway. He had been the Salmo doctor, which of course meant he was everyone's doctor. I was ashamed that we had not thought to invite him, since I was quite involved in organizing that reunion. We appreciated that Sultan Minerals donated money towards that reunion.

I remember the Langleys who had about five children. One time we were driving back to camp when we saw two of those kids walking up the steep road. Dad stopped and asked if they would like a ride. They replied, "No, we're not allowed to take a ride." We drove a bit further and there were the parents playing ball with the other children. We asked how come we saw the two walking, and they replied, "They were misbehaving and we told them to get out of the car and walk for two miles."

It was an interesting place, for it was a community of people without anyone very rich or very poor. Every man was an employee. We didn't know the concept of "being poor." It was a neat place to grow up. They showed movies at the mine and we girls took ballet and tap dancing lessons, not to mention swimming lessons and a lot of other things.

Ralph Stevens was the fire chief up on the hill. His wife was much involved in developing the Salmo golf course. His daughter Judy lives in Vernon and is married to Mark Wakefield who was the fire chief there. Does firefighting run in the family?

In Krestova, they used to call Al Nord "The Englishman." That is so funny because his parents were Norwegian. I also remember hearing about Hugo Berg, who was rumoured to make the most bonus.

ERNIE STENZEL

We moved to the Emerald when I was about three. And I stayed there until I was grown up. My dad, Frank Stenzel, died when I was eleven, and Mom took over as head of the family. Canex was an excellent company that looked after its people. An old man named Mel Olson—he was old when I was a kid, *[he was actually in his 30s then]*—he got my mother Mary Stenzel a job. He was a wonderful man, the money man, the chief accountant. He called in my mother and asked, "What are you going to do, Mary?" She said, "Well, I don't really know." Mel said, "You are still young (I think she was in her early forties). Why don't you go to Pitman's Business College and take some secretarial and bookkeeping training, then come back up here and we'll hire you? Meanwhile don't worry about the rent on your house."

Ernie Stenzel, brother of Elaine, lives in Prince Rupert where he still works with the hammer he has been wielding since age one. 2007 photo.

So Mom commuted to school in Trail for a few months and when she was done, Mel hired her. Mel had a reputation for being picky, but Mom really enjoyed working for him.

"Shorty" Kinakin worked in the mine as a timberman's helper and later on the labour crew. He and "Shorty" Mahonen did a little bit of everything. Harry Kinakin was Shorty's brother, and his wife, Miss Tickner, became my grade 4 to 6 school teacher.

Ernie Stenzel at age one. He has already chosen his life's work—swinging a hammer. Courtesy of Ernie Stenzel.

We pestered "Meat Nose" (Floyd Wilson) and "Cracker Jack" Lloyd. Glen Hartland, Dale Clayton, Arnie Anderson, Richard Miller, Alan Robinson, Bob Adolphson, and I—we pestered them for decades. If I wind up going to Hell, they will be at the door ushering us in. It was not good. We hadn't enough to do, so we used them as somebody to have fun with and drove them crazy. It was terrible. Lloyd used to park right in front facing the house. We were all hiding in the snow. Jack Lloyd gets out of the car and climbs the stairs. We wait, and then we tie a rope to his bumper and the other end of it to a tree, cover it all up with snow and then hide. Then we'd hide again until he backs out—and Bang! We pestered those poor guys and couldn't think of enough stupid things to do to them. We were a real bunch of little assholes.

In order to communicate with Lewis Hartland, who was deaf, everybody had to have a descriptive nickname, e.g. Jack Lloyd who was a bit hunch-backed became "Cracker Jack Barrel Back."

My dad had been a miner until in the early fifties the doctor told him to get out from underground. So he went to work in the carpenter shop and worked with Walter Barikoff, Ed Druggie, and Alex Baturin.

I worked in the carpenter shop too after Dad died. The mine was always good at hiring the kids—especially university students—who always got jobs in the summer. It was one of the premier companies to work for in that respect. My brother Rudy replaced Jack Lloyd and Floyd Wilson when they were on holidays.

We boys used to hang around the telephone booth behind the garage. One day this tourist drives up and says, "Can you guys tell me which is the way out of here?" Dale Clayton puts this dumb look on his face and replies, "There's a way out?" This guy just put the car in gear and took off.

Charlie Bing regularly delivers fresh vegetables to the mine. Courtesy of Ernie Stenzel.

The company opened the pool in the spring, and in June when school was out they hired a lifeguard and we had swimming lessons for two full months. Every kid could go and have swimming lessons. Every one of us had the "mile swim" ticket. Carol Anderson [McLean] gave swimming lesson at the mine and got green hair from all the chlorine [copper sulphate?] in the pool. She was very blond and before any kid ever dyed their hair red, green, or blue, she had green hair from all time she spent in the pool.

Johnny Madison was a one-armed welder at the mine. Like Adin Tetz, he had just one arm. During childhood he got his arm cut off by a train at

Blairmore. He invented the carbon arc torch because he wanted to sculpt a miner's head out of a solid block of steel. He carved two miner's heads with hard hats on them in the underground shop at Canex. One of them is in the Chamber of Mines in Nelson.

I think someone stole the carbon arc idea from him and patented it. He fought for it but I don't think he got anything out of it. That was something that came out of that mine from a man's ingenuity. They were beautiful heads, one a little smaller than the other.

Our teacher, Miss Tickner, hated when we made noise and she would often throw the brush at the blackboard to startle us, to get our attention and get us to shut up. On the one occasion she accidentally hit the principal's daughter. We also had a teacher who was later nailed for molesting kids.

It was some 40 years later when I was at a "booze camp" run by a guy in Prince Rupert, when in walks Bill Kinakin my teacher from grades 4, 5, and 6. I recognized him mostly from his hairdo and his Russian accent. I knew he had a deep voice because he used to sing for us. I went and sat down beside him and asked, "Where are from?"—"Oh, I just came from Nelson," he replied. It clicked, I knew then with certainty who he was and asked, "What are you doing up here?"—"I'm going to be a barber at Central Barbers," he said. He then asked who I was and what I did, and I said, "My name is Ernie and I'm kind of a soothsayer."—"What do you mean?"—"Oh, I predict the future and I read palms. Give me your palm so I can read it." I looked at it for a minute and said, "You are a barber, but it looks like you may have worked with your hands, maybe an electrician or a plumber." I knew he had been an electrician. "I was an electrician," he responded. "You also have intelligent fingers—you have had a good education, like a high school or elementary teacher or something" I continued. "I was an elementary school teacher," he blurted. "And you have two daughters," I added. "Holy Cow! This guy's good," he said, "Holy Cow!"—"And one of them is named Elaine," I interjected.

Dale Clayton, Arnie Anderson, and Richard ?. Courtesy of Ernie Stenzel.

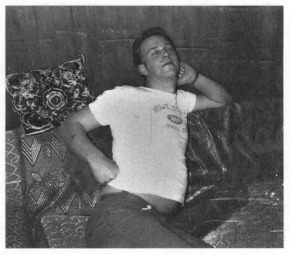

Bob Adolphson. Courtesy of Ernie Stenzel.

"Who the hell are you?" he demanded. "I'm Ernie Stenzel."—"Ernie Stenzel?" he yelled as the whole place broke into a roar.

Standing up beside the 3800 Jersey portal there was one of those worn-out 966 loader tires. There was a clear space in front of it to the road edge. We boys decided to try to move it. Now you may remember that the road from the portal went down to the compressor house and from there switchbacked down to the warehouse, then another switchback to the Tungsten portal. We rolled the tire over the bank and it went down, hit the first switchback, and bounced fifty, sixty feet in the air. It landed directly on the second switchback from where it bounced 200 feet in the air and landed in the basement of the abandoned bunkhouse. I wonder what the people who tore down that building years later thought when they found that tire in the bunkhouse.

Mickey McBlain, electrician, with his beetle. Courtesy of Ernie Stenzel.

After doing that, we went home and kept our mouths shut. We must have been fourteen to fifteen years old at the time. It never occurred to us until later that the tire could have hit someone's vehicle or done some other damage. Like typical teenagers looking for something to do, we just weren't thinking.

We did another stupid thing in the late winter. We would go down to the abandoned aerial tramway, grab the cable, and hand over hand proceed out over the drop-off, from where we would allow ourselves to drop down into the deep snow that had built up far below. It didn't occur to us that we might land on a stump and kill ourselves.

The company bought a brand-new snow blower. Ralph Stevens and Walter Barikoff said, "Come on, Ernie, we have a job for you." We grabbed the power wagon, threw the snow blower in back, and drove down to the tungsten mill. There were three to four feet of snow on all those mill roofs. They said, "Ernie, you've got to take the snow off those roofs." It was late winter and they were worried about the weight caving in the roofs.

The blower would throw the snow only about eight feet, so I started at the edges and blew it off.

Ernie Stenzel and Bub. Courtesy of Ernie Stenzel.

172 / JEWEL OF THE KOOTENAYS

Then I would make a few passes and throw it off in two stages, then later still, I needed three stages, etc. I was on that job for what seemed a month. By the time I finished, it was melting faster than I was getting it off. I wonder if they put me up there to get rid of Ernie for a while.

Later when they were rebuilding the mill, I worked helping with that too. We put in new foundations for a ball mill and I worked with the carpenters redoing these giant shaker tables that were about 12 feet long by 7 feet wide. They had these riffle strips on them that I had to take off. Then I had to put on "battleship linoleum" and with tiny nails install new riffle strips. That was my job for what seemed a hundred years.

Another job I had during my last two years of school and one year afterwards was to build ladders for underground. I had to make 25 sixteen-foot-ladders, each with sixteen rungs, every day. Walter Barikoff would cut out the dados for the rungs and I would nail in the rungs. I got to the point that with a two-and-a-half-inch nail, I would hit it twice—a tap to start it, and a single hard blow to drive it in completely—without ever hitting my thumb, because I got Walter Barikoff to get me a twenty-pound hammer. *[Ernie must have meant a two-pound hammer. Mine management probably gave Ernie a lot of work to help his mother cope with raising a family on a secretary's salary].*

JOHN BLONDEAU

John Blondeau in 2007.

I met John at Remac, where we both worked as miners, in the fall of 1952. He became a good friend and has remained one ever since. He now lives in the Riondel area.

I was born in Saskatchewan in 1932 and began to work underground at age 15 in a coal mine near Estevan and stayed there for two winters.

After working at the mine at Field, BC, I next worked at Remac. When that mine closed in the spring of 1953 because zinc prices collapsed, I moved to the Emerald mine. While working there I shared living quarters with Alex Dychuk and Wally Panagopka in Salmo and commuted to the mine by

Miner's chair, a.k.a. CIL powder box. These wooden boxes, with their dovetailed construction, could be found in every miner's home. They were used as chairs and storage containers, etc. Laid on their sides and stacked on top of each other, they made great shelves in the basement. When you visited a miner, instead of saying, "Pull up a chair," he would likely say, "Pull up a powder box," even if he meant a chair. It was a sad day for the miners when these wooden boxes were replaced by corrugated cardboard ones. Courtesy of the Salmo Museum.

company bus. I was working in the −32° winze, but on a different shift, when Adin Tetz lost his arm there. I was at the mine too when Trachsel was sucked down an ore-pass and buried, but miraculously survived.

After a short stint in Kemano I returned to the Emerald, but in 1955 I moved on to the HB mine. By the time I retired, I had spent over 50 years in underground mines. I wonder how many men can equal that record. Unlike many others, the Emerald Mine accounted for a very small fraction of my mining career.

THE TOM SMITH FAMILY

I was able to interview Enid Smith and her daughter Deb by telephone. They lived at the mine from 1954 to 1967. Enid's daughter, Carol now lives in England with her husband, while her son Peter and his wife live in Taiwan. Enid says of her son, "Peter couldn't settle down for the longest time and has travelled all over the world."

Daughter Deb is married to Steve McKinnon. They had two daughters as well as a son who is deceased. She lives near her mother in Cowichan Bay, Vancouver Island, where Enid, at age 87, still sings in choirs. Deb is a singer-song writer with three recordings to her name. She is also the co-founder of Cowichan Folk Guild and of the Island Folk Festival, Duncan, BC, in 1985. Her song "Iron Mountain" is reprinted here.

Here is what Enid Smith had to say.

The Smith Family: Enid, Carol, Deb, and Tom with Peter in front. Courtesy of Deb Maike, née Smith.

I got a "London" degree in music through the University of Manitoba and gave violin lessons at the mine. Fifteen years ago I tore the tendons in my bowing arm and have had to give up playing. I still used to tune the instrument once a month, but finally conceded that I would never play it again, so I donated my viola to the Conservatory in Victoria. I now live in Duncan, BC, near my younger daughter Deb.

My husband Tom was born at Crystal City near Winnipeg, Manitoba, and grew up on a farm. He was in the service during the war and graduated as a geologist from University of Manitoba in 1947. I was born at Buchanan, Saskatchewan, in 1920. We moved up to the Emerald in 1954, where Tom was the chief geologist in the lead-zinc mine, and we stayed there until 1967.

IRON MOUNTAIN

At four thousand feet in the heart of the Kootenays
On the side of iron Mountain our spirits ran free
Below our swift feet our fathers were picking
The ore from the mountain, their life's wealth, their reason to be

Our fathers were miners of lead, zinc and tungsten
A whistle would warn us of blasting below
Great chunks of iron ore put the food on our tables
The wind in the trees and Lost Creek, put love in our souls

> Do you ever go back to 'The Hill' as we called it,
> To the concrete remains of the old swimming pool?
> It's all that is left on the side of Iron Mountain
> The roads and the playground of Harold Lakes School

We were the dryads, the nymphs of the mountain
The Ridge's wild magic can never be known
By those in the city with concrete surroundings
And no virgin forest and no cliffs, no place to call home

> Do you ever go back to 'The Hill' as we called it,
> To the concrete remains of the old swimming pool?
> It's all that is left on the side of Iron Mountain
> The roads and the playground of Harold Lakes School

Our souls' earthly magic is all that can bind us
Our houses have vanished, the miners all gone
A tracing of roadway still winds up the mountain
To the land of the black bear, a wild world, our spirits call home

Deb Maike
SOCAN (1986)

Used with permission

JEWEL OF THE KOOTENAYS / 177

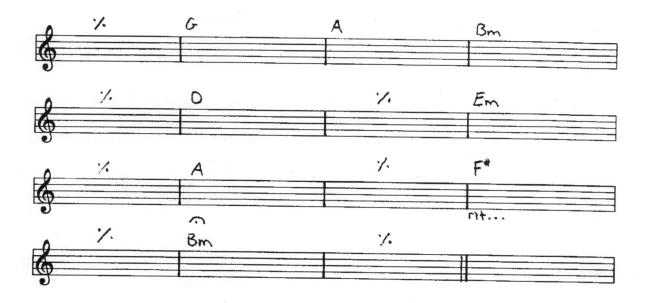

Our younger daughter Deborah was born in 1953, and our son Peter in 1956, while we lived at the mine. We were in a duplex "cracker box" to start with. Then we moved into a Panabode and I absolutely loved it there. We had a little wood stove in the centre of the house and it kept us warm. It was a lovely place to live, and we took lots of walks.

When my husband Tom transferred into Exploration we moved to Sydney, Australia, where we lived for a year before moving to Papua New Guinea, where we stayed until 1971, when we returned to Vancouver. Tom got cancer a few years later and passed away in 1981. He was such a nice man.

I recently had a slight stroke so have not been feeling well lately.

Deborah Maike told me a little about growing up at the mine.

I was born just before we moved to the Emerald and lived there until I was thirteen. My father was Tom Smith, a geologist at the mine, and my mother's name is Enid. After leaving the Emerald, the company moved us to Australia, New Guinea, and back to head office in Vancouver. Both my mother and I now live in Duncan, BC.

The one thing about the mine I shall never forget was our heated swimming pool. Kids from Salmo and the HB mine used to come up to swim in it, for they had nothing but the river down below.

We all learned to swim and I became like a fish. We had a trainer/lifeguard called Judy Anderson who put together a swim team to compete with other teams in the region. I won a lot of medals and set a provincial record for the under-eight-year-olds when I swam 25 yards in 16 seconds.

Lost Creek has also stayed in my memory, for we children all fished there when we weren't swimming. Our fishing was probably more of the Tom Sawyer/Huckleberry Finn kind with a pole and a piece of string with a hook tied to it.

I especially remember Adam Kowalyshyn, one teacher we had at the mine. He now is retired and lives in Nelson, BC.

Mr. O'Connell was our neighbour at Canex. He was a miner, and his daughter Eileen is my age and now lives in Chilliwack. His son Ken is a teacher and principal, who lives in Duncan.

AL NORD

I knew Al Nord's father slightly, for he had a garage in Ymir back in the mid fifties. I remember Al as a teenager driving a hot rod he had built. I ran into Al by coincidence around 2002 when I stopped at a diner at the South Slocan junction. When we started chatting, we soon realized that we knew each other from a half-century earlier. Al Nord's photos can be found throughout this book. He worked at Canex from 1953 to 1973, mostly in the crusher and conveying system.

I was born in the old Nelson hospital in 1936 and lived in Ymir as a young kid but moved to Zincton in 1945 when I was 9. My dad was a master mechanic at the Lucky Jim, and when he went into the TB sanatorium at Tranquille in 1949, in what is now a part of Kamloops, my brother took over from him.

I worked evenings at the mine during the wintertime when I was about thirteen. I was still going to school, and my job was contract-painting the cookhouse interior after 6:00 p.m. when the place was empty. I first washed down all the walls and then I painted them after they were dry. When I finally got a steady job with the company, it was in the mill at miner's pay—about $9.27 per day.

Doehle was the big wheel there, and his daughter married Jack MacIntosh, who was the mine superintendent and later managed the Paradise mine which Sheep Creek Gold also owned. After the Lucky Jim shut down in 1953, many of the men went to the Mineral King, another Sheep Creek property west of Invermere.

When I learned that the mine was closing, I immediately got a job with the CPR. I thought the rail company was primitive as well as

Al Nord, circa 2003.

cheap. I was paid $1.08 per hour and there were no coffee breaks. You could count on being shorted on your overtime hours. Thankfully I was not there very long, before I got a call from National Diamond Drilling to work for them at the Victory Tungsten near Salmo. It was up Sheep Creek and across from the HB mine.

We drilled there until the snow shut us down that fall. I then went to work for Connors Diamond Drilling up Lost Creek until that contract ran out, but I got another job with them drilling at the Jackpot mine on Porcupine Creek near Ymir.

My friend and I next went up to the Emerald, where his dad was working. They needed one man and since I was seventeen and he wasn't quite, I got the job working with his dad. They told me the job was for only a short time, but I stayed there until February 1972, almost 18 years. I was a shift boss in the mill when I finally left.

I worked in the crushing and conveyor system during most of my years at the mine. We had an awful lot of bearings to maintain on those conveyors. There was a set of three rollers every four feet and each roller had two bearings. Since the conveyors totalled 7,150 feet, there were about 5,400

Al Nord's house—his Skidoo appears popular with the children. Courtesy of Al Nord.

Al Nord cleaning up a "blow-out" spill with a slusher hoist. Courtesy of Al Nord.

Cleaning up a "blow-out" spill with a front-end loader. Courtesy of Al Nord.

rollers and 10,800 bearings in the system, as well as over 14,300 feet of belts.

It takes every type of man to keep a mine going. One kind of man can't do it all. I couldn't stand working in the mill where I could predict what I would be doing 10 years from now at 4:00 p.m. Nothing changes unless a lightning bolt strikes the power line and blows up a motor. Then you get some excitement. It is the same with breakdowns. When I was a shifter in the mill, I wasn't allowed to do any mechanical work—wasn't allowed to touch anything. If I wanted something built, I designed it and sent it up to the machine shop and let them build it. Since I like to build things, it was not the place for me.

I started at the mine as a "grunt," worked one day and got a raise of $.02 per hour because the union had just signed a new contract. Although I got a raise, another guy lost a cent per hour due to that same labour agreement. We went on strike one year for six bloody weeks and wound up with a five cent increase, and out of that increase the Salmo boys lost their free bus ride to the mine. We were asking for 10 cents per hour and I asked, "What happens if they offer us 5 cents?" I was told, "Oh no, we won't go back until we get the full 10 cents." Frank Clayton and I were over at Merritt at the time when we heard they had settled for 5 cents. We could have had that without a strike and without being out of work for six weeks. We could have worked there for the rest of our lives and never made up the difference in the pay we lost.

After the Emerald shut down, I worked in the east Kootenays for a while and was on the union executive when they had a six-month strike. I left and went to work on the green-chain at a sawmill, where I had to handle the 22 foot long 3"-x-12" special planks they were cutting for the CPR. After about two days of that I was almost dead. A guy there said to me, "Christ, Nord, I will have to show you how to handle

Typical house—note the other houses, almost hidden from each other. Courtesy of Al Nord.

them." He grabbed a plank by the one end and flipped it! He said, "You just have to know how to do it." Once I caught on, there was nothing to it, even though they weighed up to 180 pounds each. It had been a similar lesson when I had first begun driving spikes for the CPR. You were required to stand on the same side of the rail as the spike you were driving, but I always stood on the opposite side if no supervisor was around, and was able to drive a spike with three easy blows. In the mine we used to carry drills that weighed about 167 pounds and place them on the jumbos with no trouble at all. I weighed only 132 pounds then, but thought nothing of it whereas today, I would have trouble packing the air the machine used.

After I left the mine I worked on paving work until I retired. Now I build things all the time, and when I want to build something else and need material, I dismantle what I had built earlier and reuse the old stuff. It gives me something to do.

JOHN (JACK) ROBINSON

I don't recall ever having met Jack Robinson, who came to Canadian Exploration Limited in 1954, but I have certainly become familiar with his name. Jack went on to become a Mines Inspector for the BC Department of Mines, a job he held for a number of years. Cliff Rennie told me that Jack is now in a seniors' home in Nanaimo. The following is based on an article in The Jersey News *(April 1969) by Mel Olson.*

Jack started as a miner in the Emerald Tungsten, worked as a surveyor, and then with the geology department for about one year. He was made Tungsten mine engineer during the sinking of the Emerald shaft, and later

Jack Robinson as Santa in his sleigh (built by Al Nord). Courtesy of Al Nord.

became a shift boss in the Dodger mine. He was project engineer, Jersey mine engineer, and was the Jersey mine superintendent since January 1961 until he left.

Jack was active in local clubs and held executive positions in many associations. Some of the ones he will admit to were president of the Emerald Community League and vice president, etc. in the Emerald Curling Club. He was President of the Nelson Toastmaster's Club and an excellent speaker with a Dale Carnegie "enthusiasm" background. He was active in mine rescue, first aid (7th Label), and safety, having taken safety for supervisors and many similar courses. He was also a member of the Ymir Masonic Lodge.

Jack graduated from the University of British Columbia with a B.Sc., and was a member of the Canadian Institute of Mining and Metallurgy and the BC Association of Professional Engineers.

The Jersey Mine had an enviable record under his supervision. Production increased steadily from 30,000 tons to 50,000 tons per month, and he was enthusiastic about the future and of every man and machine in his department.

Records indicate that Jack gleaned a wealth of experience from years of practical experience at mines that included Bralorne, Yukon Consolidated Gold Corporation, Highland-Bell. Kootenay Belle, Yale Lead and Zinc, and many others. He also worked hard to operate his own rock quarry near the Black Bluffs beside the Nelson–Nelway highway, which had been a venture offering him profit, business experience, and fresh air.

During his 15 years at Canadian Exploration, Jack was an enthusiastic swimmer in the summer and curler in the winter. He was also active in hockey, shovelling snow off the rink roof, working on other community projects, and was a past president of the Emerald Community League. When he came to Canadian Exploration, he had two small children. They were nearly grown when he left.

JOAN HALLBAUER

I never knew the Hallbauers, but Joan's husband Bob was so well-known that it was impossible not to know about him if you were in mining. Perhaps Joan got spoiled at the Emerald where she probably had a million dollar view. She recently moved out of her house and into a penthouse in West Vancouver near the waterfront—and again has a wonderful view.

We came to Salmo in 1954, but because of a threatened strike, the mine hired Bob on as a miner instead of an engineer. He had, after all, grown up at the Sheep Creek Gold mine and had worked underground there during his school years. In addition to Sheep Creek, he had also worked at Riondel, Zincton, and the Mineral King mines during the summers. Besides working at Sheep Creek, Bob's father had worked at the Paradise mine for several years.

We stayed in Nelson, where I lived with my mother until the company found us a house. It was in the new townsite where the panabodes were, but ours was one of the cracker box duplexes—20 feet wide by 24 feet long per side. It had two bedrooms, a living room and kitchen and was built of plywood. You could easily hear the people talking next door. Just the same, we were very happy. You can still find some of these old houses being used as part of a motel on the Lower Nicola highway near Merritt.

Joan and Bob Hallbauer, early 1950s. Courtesy of Joan Hallbauer.

I was born in Nelson, as was my mother—who was born there in 1895 and lived to be 98. She passed away just two years before my husband, who died in 1995. I did my schooling there and took stenographer's training as well. Then I went into nursing and trained at St. Paul's Hospital in Vancouver.

Bob took his high school in Nelson as well. He was born in Nakusp and raised at Sheep Creek Gold mines, where he got his elementary schooling in a little one-room schoolhouse. He finished his senior matriculation in Nelson and his mining degree at UBC.

In 1952 we were married in Nelson while Bob was still at UBC. I worked for a year and a half before I got pregnant, and our first son, Russ, was born in January 1954 shortly before Bob graduated in March. Our second son, Tom, and our daughter Cathy were both born in Nelson after we had moved to the Emerald. It seemed to me that I was

The first two of Hallbauers' three children, Russ and Kathy, at play. Mid-1950s. Courtesy of Joan Hallbauer.

Joan's rock garden, early 1950s. What else can she grow? Courtesy of Joan Hallbauer.

It seems that even the grown-ups are not too old for a toboggan ride. Mid-1950s. Courtesy of Joan Hallbauer.

pregnant all the time while we lived up there. My story was, I married a mining engineer to see the world and instead I moved to Salmo! What a disappointment at first, but we had a wonderful time. Bob worked as a miner for only three months. When the threatened strike blew over, he went into the engineering office where he started at the bottom. The company believed in training its people for new opportunities. That way when they opened a new mine, they always had a core crew ready. I never worked (paid work that is) after I got pregnant, isn't that awful?

We never owned a washing machine until my third child was over a year old. I washed diapers and clothes every night in the bathtub and hung them around the hot water tank to dry. We had a lovely clothesline outside too.

At first we couldn't afford a washing machine. Then I got stubborn—I wanted an automatic— they had just come into the stores, but Bob didn't have enough money, so I waited until he could afford one. Before that happened, Marie and Joe Adie gave me an old copper job that they had in their basement. Joe was the chief geologist there at the time, and he passed away a year or two ago. He gave us this huge, great big old copper washer with a wringer and that was a godsend for me. Bob would lug it into the kitchen and fill it with hot water, but I had to turn the agitator in it by hand.

Grandpa came up one day and wondered what I was doing, agitating the machine by hand. What's wrong with my son? Why doesn't he put a cotter pin in it? So he drilled a hole through the agitator and the shaft and put a nail through it. That worked. The agitator turned on its own after that. It must have been an expensive washer in its day for it was solid copper. I loved it because the wringer worked as well. I must have a very bad memory, for everything I remember seemed good.

At the mine it was not all work and no play. We had a small musical band and we had parties at Halloween and at Christmas for the kids and curling bonspiel parties too. It was wonderful! Absolutely wonderful! Bob, Tony Triggs, Garth Jones, and another fellow used to participate in the Salmo bonspiels. They were also very good bridge players, but I was never any good at cards. I was not interested and I had no card sense at all. Bob was an excellent cardplayer and he could remember hands he had held way back when. I did play bridge, but poorly. I had to have all the rules written out in front of me—no retentive memory at all. I've always been like that but Bob had a wonderful memory, one that I never ever questioned.

It seemed that everybody living at the mine had children. It was wonderful—I have such good memories of that time. We didn't have very much, but the pay was good. Bob started off at $310 per month and we paid $0.25 a day for rent, but we did have to pay for our heat as well. It was a wonderful life. I've still got copies of our utility bills.

I will never forget the New Year Eve's party the first year we were there. It was in the community hall near the manager's house. Everyone was dressed up in suits and gowns and we were all pretty high. The bathroom in the community hall was on the second floor accessed by a long exterior flight of stairs. On my way back down, my good friend Nora kept urging me with, "Come on Joan, get a move on." I lost my balance and fell. I tumbled ass over tea kettle all the way down the stairs, from the top to the very bottom. The following day we had to go to Trail for a dinner party with Bob's father and uncle. I was not only hung over, but I was in such dreadful pain that they finally took me to the hospital. I had broken the knob off my elbow! I couldn't curl for two years after that because I couldn't sweep. Every time since then when I have seen my friend Nora Triggs, I ask her if she remembers the night she pushed me down the stairs. Of course she did no such thing, but just the same, I always give her a bad time about it.

We had one of the first TV sets at the mine. We were in Wosk's store in Vancouver, I don't remember why, and bought one. At home we had to put an antenna on the roof and try to get it turned just right. Whenever we had a big wind, the antenna would sometimes turn a little and we got to look at a lot of snow. We often saw our neighbours up on their roofs adjusting their antennas too. The only channel we could get was CBS from Spokane, and we regularly watched Lucille Ball and Desi Arnez in "I Love Lucy" as well as other programs. The TV, however, changed community life. When we visited a neighbour, conversation would be limited to, "Hello,"—"Do you take cream and sugar?"—"Hush," and "Good night." I guess that's just about what happened all over the country, not just up there. Nevertheless I don't know what we did without it.

What else did we do up there? We had lots of coffee klatches and time with the children, who all played together. We women tended to be indoors all winter, but of course the kids got out and played in the snow.

Talk about snow, we really got it up there! One year in November it seemed that the whole province froze up—froze up solid. I don't remember the year, but I think I was pregnant with Kathy and that would have been in 1955/56. Almost every house in Salmo was frozen too. Bob was away hunting at Creston with Tony Triggs and a couple of other guys. My mom and dad were gone, and while Bob was away I stayed at their house to keep their coal and wood stove going. When Bob came back up to the mine, every toilet in the whole townsite was frozen.

We used to go into Nelson to buy our groceries every two weeks. It seemed to be such a long trip in those days, for it was 34 miles of winding highway. It was not well maintained and we were driving a little old Austin. We used to take our old tires in and get them retreaded with sawdust mixed into the rubber for winter driving. When I told this to my son, he said, "Oh Mother! Come on!"

Bob moved to the Craigmont mine at Merritt in 1959, but I didn't move until early in 1960. We left Craigmont in 1968 and went to Teck Mining in Toronto, where we stayed for two years. Teck then took over Cominco. It was like a mouse swallowing an elephant. Cominco had quite a crew in those days, very top-heavy. At Kimberley they actually paid men to not work when Bob took over. He found an unbelievable amount of feather bedding. He wound up with two big strikes, one in Kimberley and one in Trail, but they had to pay the price or go down the tube.

I have always told my children that their father was a producer. He liked to build a mine and make it work. Bob sometimes got discouraged because it took over ten years to get a new mine into production. He'd say, "You have to keep all the troops happy while dealing with naysayers as well. You sometimes begin to wonder why people even try. To be a miner you have to be a dreamer." He thought he would be retired before the Polaris or Red Dog mines got into production, but he did live to see it. He often wondered if a mine he was building would ever produce anything. My 53-year-old son Russ is with Taseko, and he is like that too. He has a couple of wonderful projects on the go, but wonders if he will be retired before they go into production.

Russ was with Cominco for 22 years as a mining engineer. He was at Tumbler Ridge for 15 years with Teck. Now he runs Taseko, which owns the Gibraltar mine and has some other properties as well. My second son was a "mining tech" and has been at Smithers, Houston, and Stewart before coming down to Fernie. He has quit mining and is now a farmer. He was always a farmer at heart. He didn't like shift work because he was a poor sleeper, and I'm so glad that he changed. I hated shift work too when I was nursing. It is not normal.

My years at the Emerald are unforgettable. If I tend to remember only the good parts of it, that is okay, because they were the most memorable years of my life.

ROBERT E. HALLBAUER
1930 – 1995[14]

For almost three decades the 1970s, 1980s and until his death in 1995 Robert Hallbauer was recognized by industry, government and labor as a giant in terms of his presence and influence over the mining industry in British Columbia. He came to personify the industry, to symbolize it, not by design, but by the strength of his character, his integrity and his technical accomplishments.

Hallbauer was a builder at heart, best known for developing a series of mines that led to a new period of growth for associated companies Teck and Cominco.

Born into a West Kootenay pioneer family in 1930, Hallbauer graduated from the University of BC in 1954 with a B.A.Sc. in mining engineering. He joined Placer Development as an underground miner at the Emerald mine in Salmo, and progressed quickly through the organization, rising to the position of manager at the Craigmont copper mine in Merritt.

In 1968, Hallbauer joined Teck as vice president of mining. Over the next dozen years, he spearheaded the construction and operation of many mines. In 1975, the Newfoundland Zinc mine at Daniel's Harbour started production on time and budget. The following year, production started at the Niobec mine in Quebec. Hallbauer was the first senior manager at Teck to recognize the potential of the Afton mine near Kamloops, BC. He urged Teck to "go after it", even though some felt Afton's native copper content might pose problems. Afton turned out to be a major success story after it was placed into production in 1978.

In 1981, the Highmont copper-molybdenum mine began operations in BC. Its mill assets were later merged into the Highland Valley Copper complex, which today is the second largest copper concentrating facility in the world. Another of Hallbauer's contributions was his involvement in the Northeast Coal project, which made its first shipment to Japan in early 1984. And under Hallbauer's direction, the David Bell gold mine at Hemlo, Ont., opened on time and poured its first gold in May of 1985.

Hallbauer was a man of vision who played a key role in the evolution of Teck from a modest producer to one of the world's largest mining companies. He also helped rebuild Cominco after a consortium

Robert Hallbauer. Courtesy of Kea Barker, Teck-Cominco.

---

14 *Reproduced courtesy of The Canadian Mining Hall of Fame's website,* wwwlhalloffame.mining.ca/halloffame

headed by Teck took control of the debt-ridden company in 1986. The 1987 decision to develop the Red Dog zinc-lead mine in Alaska, thus assuring the future of Cominco and its smelter at Trail, was another of Hallbauer's undertakings. In more recent years, he oversaw the construction of the Snip gold mine in BC, and the Quebrada Blanca copper mine in Chile. He also modernized the Trail smelter complex, and fast-tracked the Kudz Ze Kayah zinc deposit, a Cominco discovery being readied for production in the Yukon.

For his many accomplishments, Hallbauer was one of two recipients of The Northern Miner's "Man of the Year" award in 1982. In 1992, he received the CIM's Inco Medal.

Hallbauer was an outstanding spokesman for the mining industry, having served many years as a director of both the Mining Association of BC and the Coal Association. He was an industry leader, with a well earned reputation as a determined champion of mining interests in both public and private forums.

THE TETZ BOYS

I did not know Adin Tetz, but I remember seeing him around Ymir and I knew who he was. In my research for this book I spoke with his niece Sharon on the phone and interviewed two of his six sons in person. Following is some material generated from those conversations as it pertains to these two sons. The material pertaining to Adin's accident is in the next section, "The Dark Side of Mining."

TERRY TETZ

I lived and worked at the Emerald mine for six years, much of it in the underground crushing plant with Al Nord. He was a great guy, and he was also our "great white hunter." If we wanted to get a few extra minutes at the end of our lunch break, one of us would say to Al, "Tell us about the time when you went hunting at . . ." And away he'd go. We would sometimes get an extra fifteen minutes while he finished a story. Al was a Ski-Doo freak too, so that was another good topic at lunch time.

I was born on July 16, 1947, and grew up in Ymir. I just turned 60 a few days ago and worked at Canex for six of those years—they were six wonderful years. That job stuck with me for a long time. I was married and we already had a baby before I went to work there.

By the time I left the Emerald I knew those conveyor belts from one end of them to the other. Between lubricating the thousands of rollers, repairing the belts, and cleaning up the spills I got to know it better than I ever wanted to.

After the company went back into the scheelite [*tungsten*] ore, I went into the lead-zinc mill on maintenance and later as a mill operator. I worked with Dennis Hartland there and liked him a lot. We became really close and drank a lot of beer together, but never at work.

I did not have Bonnie Klovance as a teacher, but knew her quite well because I often sat with them in the bar. I don't know how I could have forgotten about the time when they pulled Gil Mosses out of the chute. It was amazing—them getting him out of there.

Growing up at Ymir and going to school in Salmo was a great experience, for I got to know everybody. I liked the people I worked with, they were great. I will never forget John Bishop. He was a wonderful guy and the same age as me.

We had a bunch of Nova Scotians in Ymir who all came from Walton and married the available Ymir and Salmo gals. One thing about those guys was that if they scrapped amongst each other, you had best not interfere, for then they would forget their differences and all turn on you.

At the mine we were all part of each other's family. I remember Joel Ackert from Salmo. He was a great guy and had the best collection of mining stuff around, even after giving much of it to museums. He is probably the best collector around that I know. I liked Lawrence Bond. After he was a little too old to mine, he came down and worked in the crusher with us. He was a wonderful man.

The mine provided a wonderful way of life for a lot of people. These were all guys that I drank a lot of beer with.

Terry Tetz grew up in Ymir, but lived and worked at Canex during its last years. His father was there for many years too and left his left arm there. 2007 photo.

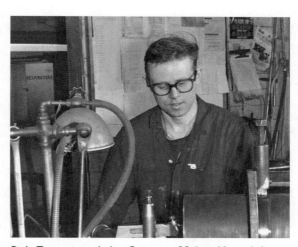
Dale Tetz using a lathe. Courtesy of Sultan Minerals Inc.

GARNET TETZ

I was born in 1953 and grew up in Ymir, but my older brother Terry worked for the company and lived up at the mine too. He became a millwright there. My dad became a welder after losing his arm and worked at Canex in the machine shop in that capacity.

As teenagers, in the mid-sixties, we used to go up to hang out at the mine, and I remember coming back down the hill once. We made it, but our friends in the other car didn't. They put it over the bank and got smashed up pretty good. We sometimes went to Rosebud and to Erie Lakes for picnics. I remember that after Dave Martin painted the Ymir hotel, we called it the Purple Peanut.

One time my friends and I were at my brother Terry's house, waiting for him to come home from work. I was about 17 then, and we were upstairs having a beer. We heard this noise in the basement. "What's that down there smashing around? It's a burglar. Let's go down and get him." So we ran through that door all gung-ho, we were going to kick the shit out of the guy.

It was a bear. We looked at it and it looked at us and we ran for the door, but the bear ran right over the top of us knocking us to the ground. We got up still running. We didn't even realize the bear had run right over us until we saw it in front of us, still going flat out.

PAUL AND PAT GRETCHEN

The names of the Gretchens had been familiar to me back in the fifties, although I never met them until I looked them up in their Irly Bird building supply store while doing research for this book. Paul Gretchen is recovering from a stroke, so his memory sometimes needed help from his wife Pat.

I chatted with Paul first, while Pat was looking for photos.

I started to work at the Emerald Mine in April of 1955. I was only sixteen, but told them I was nineteen. Pat and I got married on October 12, 1957, and moved to the mine the following day—into a "matchbox" duplex. I think we paid about $25 per month and heated the place with oil. When I got married, I was an underground loader operator—on an Eimco 104 overhead loader. I had hired on as a pipe fitter, but somehow that qualified me to run a loader. I also drove the Dumptor trucks.

Because I ran equipment, I was often asked to run equipment on the weekend on a volunteer basis for the community projects. The company supplied the materials and the workers provided the labour, for projects like the swimming pool, skating rink, curling rink, and the Salmo ski hill.

I finally applied for a blaster's certificate after our third move at the mine. I decided that I wanted some of that "bonus" the miners were earning, so I applied for and got mining jobs.

Pat then chimed in:

We lived in the matchboxes for a couple of years until we had Debbie, when Paul applied for and got a Panabode, so we moved to the other side of where the guest house was. The Panabodes were hard to heat because they had no insulation, but they were better than the matchboxes [*duplexes*] where there were no secrets—you could hear everything that went on, on the other side. The Panabodes were much more private. We also had a little yard where the kids could play.

We had two daughters while we lived at the mine, both born in the Nelson hospital. When they got a bit older, there was a job available in the commissary, so I applied for and got it. Mrs. Rhea Colwell, whose husband was an accountant, was in charge of it. It was my first venture into business. The job was for only a couple of hours a day and I don't remember how much I was paid. Part of the job was to do a monthly inventory report. The management told me that I was the first person in the store to make a profit. It made me feel

Pat and Paul Gretchen in their Irly Bird building supply store at Salmo, 2007.

Curling foursome. Members not identified. Courtesy of Pat and Paul Gretchen

pretty good. I think Nancy Verigin or Eileen McConnell took it over after I left. Eileen's husband was an electrician. They were the parents of Ross, who works for the village of Salmo, and John, who is now the president of Keltic Mines. Both boys have babysat our daughters.

Paul worked at the mine until the year before they closed it and auctioned everything off. We had moved into Salmo a year or two earlier. When Ritchie's came up to sell off the equipment, we already had a restaurant, and the Ritchie people all ate there. We also provided the men with lunches to take with them to the mine. We called it the Rainbow, but it is now called Charlie's. We lived in the back of the building. The main street was still part of Highway #6 then, so we got a lot of drop-in traffic.

We already had the pool hall in Salmo when John Panagopka came to us one day and said, "I have this land on which I am going to build a restaurant. I want you guys to buy it off me when it is finished." He added, "Sell your house and live in the back of the restaurant." It was kind of scary, because we didn't know much about restaurants and cooking.

When we moved, we took all the pool tables and put them in the basement below the restaurant, sold our house and moved into the back of the restaurant. We rented a U-Haul and drove to Vancouver, bought the restaurant equipment and hauled it back to Salmo. We were the first restaurant to have a liquor licence here. We had to travel all the way to Victoria to apply for that licence because we couldn't do it by mail in those days. We did some of the cooking ourselves, but used hired cooks as well. Basil Zakerdonski's wife was one of them. Paul often cooked the breakfasts and I came in to make the soups and things like that. I think we paid about $35,000 for that building. It was quite an experience, for Paul had little formal education. However, I had my high school and the experience of running the commissary. My parents used to have a store at Thrums, where I often helped out, so I did have some business experience. Furthermore, I had also worked in a grocery store for a while. On the other hand, there were seven or eight other restaurants and cafés to compete with. I used to make a lot of pies back then too, but have not baked a single pie since we sold that restaurant.

Paul then told me a little about his childhood.

I was born on a farm at Waneta. My father, who had come from Russia at a very young age, had no schooling whatsoever. We had no land and were extremely poor, so Dad worked for other farmers in the area. I remember living in a house on a farm right beside the Pend Oreille River, just above the present dam site. Something I have never forgotten was that someone gave me a homemade tricycle when I was small. The reason I got little schooling was that when I got older, I accompanied my father to work on the farms.

Then Pat finished off the conversation by telling me about her childhood.

I came to the Castlegar area at a young age. My dad had land in Saskatchewan, so we travelled back and forth like nomads. We went to Saskatchewan whenever the crops were ready to plant or harvest and lived in Thrums the rest of the time, where my parents ran a business called "The Window Store." My mother was also the local driver who hauled her neighbours to the market and back. She was paid five dollars per trip for this little chore.

I hated one thing about Paul mining—the dirty clothes he brought home for me to wash. The washing machine got so filthy from them that it was a dirty job to clean it.

We now live in Castlegar, and commute to Salmo daily to tend our Irly Bird building supply store. However, the best times of our lives were the years when we lived at the Canex Emerald Mine.

BOB DODDS

I have known Bob for a long time, even from before he married my next-door neighbour's daughter at Remac, who was our babysitter. I remember meeting his older sister, Irene, at a local swimming hole in the Salmo River as well. In the sixties he showed me how to tie fishing flies when I visited him while he lived at Merritt. He is past retirement age now but shows no signs of ending his consulting business anytime soon.

My dad started at the Emerald in 1947 in the tram shed. He used to work on Saturdays a lot. That was general maintenance day. I would sometimes go with him, and afterwards we would go fishing in the river. I believe Erickson was the mill manager then. His sons Ron and Kenny were my classmates. They and I spent a lot of time together at the mine. Back then the tailings pond amounted to nothing. I believe they may even have been running the tailings straight into the river. The river was just soup, but they cleaned it up and did a superb job of doing so.

My first (summer) job was at Sunshine Lardeau, where I roomed with a miner who taught me what not to do. We were paid every Friday in cash. I think I would get about $150, and he would get close to $1,000. He'd say, "Come on, I'll show you what we do now." He'd sit down at the table where the cards were flying, and soon he would get up and say, "Let's go." I would say, "What about your money?" He would retort, "Well, you win some and you lose some." Before the week was out he would be bumming money off me. I decided right there that gambling was not going to be a part of my life.

After I graduated from Grade 12 with senior

Bob Dodds, 2007.

Bob Dodds—an ardent fly fisherman with a big one from the Salmo River. Late 1950s. Courtesy of Ron Erickson.

matriculation, I wrote the UBC entrance exams but could not afford to go. A few years later I would go to the Heavy Duty Trade School in Nanaimo—there were two in BC at that time, and the other was in Prince George. They were both run privately even though the government funded them. A fellow named McCready ran the one in Nanaimo. He was quite a character, but it was an excellent school. Charlie Scribner from Salmo went to the school in Prince George.

At Canex I started underground as a helper with Connor's Diamond Drilling sometime in 1955 after I turned 18. I worked there for about two months before being laid off. I was then invited to have an interview at the engineering office to become a surveyor's helper. I first worked as George Sutherland's helper in the tungsten mine as well as Sawatzki's, the surveyor in the Jersey lead-zinc mine. They taught me how to calculate and plot the data.

It was not long after I came that Sawatzki left, so George and I soon looked after both mines, but in the lead-zinc mine George would be my helper. Sometimes Ross Duthie was my helper, but most of the time I worked with George Sutherland. We took turns running the instrument to cut down on the amount of walking. The two of us were soon doing all the surveying. We were fast—we would trot much of the time, and we got the jobs done!

The inclined shaft in the tungsten mine was a nasty one to survey because it had so many raises off it as well as stopes which were sometimes joined. It was ugly. We sometimes had an awful time surveying those places. We had to measure all the production of the contract miners and plot that as well. With some of the raise miners, we just had to line them up from the bottom once, but with others, we had to go up in the raise to get them back on line. I forget his name, but we had one miner who was so good that it looked as though he had cut the raise with a knife the whole distance. He was always at the hoist station at lunch time, his round drilled, loaded, and ready to blast. I always called him the "smart miner," for he did not gamble, horse around, or drink. He just put his money in the bank.

I had a hell of a time with Ross Duthie once when we had to measure a raise in the tungsten mine. He said, "You just stay down here with the instrument and I will go up the ladders with the end of the tape measure. So he went up there and everything was silent, so I yelled "Ross! Hey Ross, what are you doing?" There was no response. After a couple more yells with no response, I climbed up to see what was going on. I got up there and Ross was frozen cock stiff on the end of the ladder! There was this huge boulder sitting right on the top of the ladder. He was so petrified that I had a hell of a time getting his hands loose from the ladder rung and getting him back down the raise. I said, "Ross, what the hell is going on?" Ross blurted, "I have never been so scared in all my life." I got the instrument out of the way and I went back up and got the boulder pried off the ladder and down the raise. I came down and said to Ross, "Now do you want to go back up with the tape and measure this raise?" He said, "Fuck you!"

Jim Dodds, mill man. Courtesy of Sultan Minerals Inc.

I really enjoyed working with guys like George Sutherland, Tony Triggs, Bob Hallbauer, and the others. Bob, George, and I hunted and fished together too. We had decoys stored at Creston and spent many days hunting and fishing over there. Later, after I was transferred to Merritt and Bob Hallbauer became the manager of Craigmont, we fished together over there as well.

Hallbauer was a very clever guy. If you did your job, there was no problem, but if you were a bit of a slack ass, you didn't hang around him too long. He may not have been very popular at Kimberley after they took over Cominco and he cleaned up the dead wood there, but he was sure popular with the shareholders. He was Mining Man of the Year at least once and was later inducted into the Mining Hall of Fame.

I was laid off at the Emerald sometime in 1957 when things were slow. It was then that they sank a shaft to go down for the tungsten ore. I worked at Remac mine for the next two years and then went on to trade school in Nanaimo for a year. After graduating I started my apprenticeship with Dietrich Collins in Vancouver. I was there for only six months when I was laid off at the start of December of 1960.

I was panicky. Gwen and I were recently married and she was expecting a baby. No job and I am an apprentice. There was a major downturn in the economy at the time as well. At my wit's end, I phoned Placer and asked for the personnel department. Doug Little answered the phone. He said, "Bob, how the hell are you?" So I filled him in on what was going on. "You come down and have a coffee with me in the morning," he said.

I went down in the morning, and Tony Triggs was there and I forget who else. We had quite a bullshit session, after which Doug called me into his office and said, "How would you like to move to Salmo?" I said, "What?" He replied, "First I want to tell you, you are on my payroll right now. You have a job. Now I don't know what you will be doing. We have a house ready for you. All you have to do is get there, move into the house, and we'll find something for you to do." So I was hired and we moved back up to the mine. You can't imagine what a relief that was. I had felt under enormous pressure being laid off just before Christmas with the economy slow. *[It was Doug Little who forgave the Adin Tetz second mortgage when Adin died]*.

I had nothing to do for close to two months. I would go to the engineering office, and they would say, "No, we have nothing, but go down and help Curly Colwell down in purchasing, and the warehouse." There wasn't much to do, believe me. It was terrible. Anyway, I was being paid. Finally in the spring they asked me if I would go underground and relieve the serviceman in the underground shop. I did that and stayed there until almost the end of June. About the middle of June they asked me if I would move to the Craigmont mine at Merritt. At Canex, I had not been in the apprenticeship program, but after moving to Merritt and working in the truck shop, it resumed. I left Craigmont in April of 1964 and moved to Vancouver.

PATRICIA ROSE, NÉE WLASIUK

I interviewed Patricia in her Nelson home where she has an excellent garden. I got there just in time to help her eat the peas she had just picked from it. She, like so many others, had some excellent photographs for me to scan. As a single mother she has not had an easy time of it, but one would never know it from her looks, for she is still as attractive as ever. Her father was Steve Wlasiuk, who worked at the mine. They lived at the mine from 1955 to 1964.

I was born in Blairmore and delivered by a midwife. We moved to the Emerald mine in 1955 when I was just finishing grade two. We lived there until Dad built a place on Airport Road near Salmo around 1964, by which time I was finishing high school and was going into nurse's training at Vancouver General. Dad stayed with Canex until the mine closed. He used to be on the mine rescue team while we lived on the hill, but may have dropped out after we moved down.

During my preteen years the mine was an exciting place for me. I loved that skating rink. They brought a shack down from the mine and set it up beside the rink and put a wood heater in it. It was cozy and we could go in and get warmed up when we got cold. It was so neat. We never stayed indoors during the winter, we were always outside. To us it was the best thing in the world.

We played a lot on the roads. I remember Lewis Hartland as a young boy of four or five. He would know if a car was coming and move off the road long before we could hear it. He was quite amazing. He may have seemed slow to some, but he was really very smart. It was odd—we would learn that he was deaf, and then we didn't think about it anymore. To us he was just another kid.

We had a huge swimming pool—oh, we lived at that pool. It was the best thing they could have done for us. The pool was heated, except for when we did our tests. We had a lifeguard who gave us free swimming lessons. All the kids at the mine went for the swimming lessons every morning. We would swim all afternoon, go home for dinner, and come back and swim until eight o'clock at night. That pool is still there. It is battered, but still there.

They always dropped the temperature during the swim tests. It caused me to fail the test for my bronze medallion. I just couldn't finish the last lap of the crawl in the cold water. I passed everything but that. For the other strokes, you didn't have to put your head under, but for that one you did. I couldn't breathe. I couldn't get air. The cold water just shut me down. *[The pool was drained and cleaned once a week, then refilled with cold water. Perhaps the swim tests occurred before the cold water was displaced with heated water from the compressors.]*

The forest was our playground. We were free to roam the hills all around the camp. No one worried about us. We regularly went to the dump to watch the bears, but it never occurred to us to be afraid of them. We would hike all over the mountainside. There was a big long ridge that seemed to go on forever *[Quartzite Ridge]*. It had lots of gullies in it, but we were never afraid to roam through

Patricia Rose, née Wlasiuk, grew up at the mine where her father was a mechanic. 2007 photo.

Wlasiuk family: Carol, Mrs. Helen Wlasiuk, Steve Wlasiuk, and Patricia. Courtesy of Patricia Rose, née Wlasiuk.

Skating rink. Courtesy of Nick Smortchevsky Jr.

Canex skating rink. Courtesy of Patricia Rose, née Wlasiuk.

Skating rink—close-up. Harold Lakes School is visible in the background. Courtesy of Patricia Rose, née Wlasiuk.

them, nor were our parents concerned about us.

The neat thing was that we weren't aware of class distinction. Being miners' children, we should have been low on the totem pole, but the townsite was mixed. There were miners, electricians, engineers, and except for the really big guys, they all intermingled. We didn't feel unworthy because of coming from different backgrounds. Houses were allotted according to the family size, not to an employee's position in the company. Furthermore, there were so few kids up there that if we wanted a baseball game, everyone had to participate. One didn't have to be good to play. Everyone was always included.

At the mine I had a teacher named Mark Phillips from grades four to six. He must have been excellent, for we all felt good about school when he taught us. His wife taught grades one to three—in the junior room. We didn't feel that we were in different grades. I didn't realize then that liking school was not necessarily the norm.

Another thing about living at the mine was that it was like one big family. If I was down the road misbehaving, Mrs. So-and-So would come out and tell me to behave. It was as though we were all related. The parents took responsibility for each other's children.

The mine was a great place to live until I reached my teens. Then I became bored, for all the things that used to be fun weren't anymore. I also started grade seven in Salmo and I met lots of new kids as well.

During that year, I became the biggest klutz you can imagine, for I grew too quickly and had no coordination at all. I got top marks and I was taller than the boys, so they never asked me out. It seems that if you are a girl, you are not allowed to be both smart and tall. This all happened at the same time as I started school in Salmo.

Sheila McLean was two years younger than me and a year younger than my sister and was always in competition with her. She was very pretty and popular with the boys. She envied my sister and tried to emulate her. She was a real character. I saw her in Nelson recently, but did not recognizer her at first. However, when she turned and looked at me, she had this unique look, a Sheila look, that dates right back from her school days. I knew her instantly.

The mine was ahead of its time with the way it treated its employees. I don't remember who was in charge, but I never experienced racism or class distinction until I got to Salmo high school where there may have been a bit of it. I eventually married a ski instructor, but that did not last. After five years of it I became a single mother with two sons who I raised mostly on my own. I didn't work much in hospitals because I didn't want the shift work. I was usually in doctors' offices and spent only enough time in hospitals to maintain my nurse's qualifications. My boys were not brought up with silver spoons in their mouths, but they are both doing very well. One lives in Japan, where he teaches and does some acting. The other one is an architect in Montreal. They were born and raised in Vancouver, but once they were through university, I moved back to the Kootenays. One thing I learned was how to stretch a dollar a long way. I had to.

Patricia's birthday party. Back row: Colin Hartland, Sandra Bond, Terry Peters, Patricia, Ron Clayton, Diane Woodin. Front row: Carol Wlasiuk, Adrian Smortchevsky, Brenda Madison, Lee Rowe. Courtesy of Patricia Rose, née Wlasiuk.

My mother was raised in the Doukhobor community where they lived communally. My mom's father worked away from home in logging camps and earned only enough money for Grandma to buy things like sugar, salt, spices, and flour. She was a great gardener and always grew enough vegetables to feed us.

My dad's father would sometimes leave and be gone for a year and send no money. During those episodes his wife would have to feed and clothe eight children on her own and with help from the community.

When my parents were married, Dad was working in a mine. He had come from Ukraine at age four, grown up in Athabaska where they were destitute. They had sold all their possessions in Ukraine to come to Canada and arrived in the late summer of 1929 with nothing. I don't know how they survived, but my aunt said they even ate grass. They must also have eaten squirrels and other small animals.

My dad had no education and he had to take whatever work he could find. During the war, he had tried to enlist, but was rejected because as a child he and his sister had lost parts of their fingers when they played with detonators. He worked about nine years at the Hillcrest mine in the

Crows Nest pass, but seldom full time. When he came to Canex, he had a lot of debt from Hillcrest to pay off, but at Canex he had a house and a steady full-time job. Besides that he was now into "hard-rock" mining, which was a safer job *[than coal mining]*.

TERRY ALLEN

I met Terry and his wife Barbara, when I interviewed Terry in their high-rise apartment on the Nanaimo waterfront. Terry was a geologist, and they lived at the Emerald Mine from 1956 to 1960. Terry was kind enough to provide me with the account written by his son Matt that we have reprinted in the epilogue.

I was born in Medicine Hat. My father was in the RCMP, so we moved around a lot, but I grew up mostly in Edmonton, where we spent 10 years. We also lived for a while in Lethbridge. Dad retired in 1946. He had been frozen in the Force during the war, so he had something like 36 to 37 years of service. We came out to the coast, and I finished my high school in Maple Ridge, married Barbara, my high school sweetheart, went to UBC, and completed my geological engineering in 1954. My first job after graduation was at Tulsequah, where I spent a year and a half before getting a letter from Bob Hallbauer suggesting I apply for a job as a tungsten geologist. I did and got the job. Hallbauer had been in my class and we graduated at the same time. We shared a lot of classes although he was in mining, not geological engineering. We both dated student nurses at the same time.

Barbara and Terry Allen. Courtesy of Terry and Barbara Allen.

Hub Maxwell was an interesting man. When I arrived at the Emerald, I innocently asked him for some tips on driving in the snow. He replied, "The only tip I can give you is that you take your car keys and throw them into the nearest snow bank. When you can pick them up off the ground in the spring, that's the time to start driving again." What I didn't know then was that he and his wife had on the previous day had an accident that had wiped out the family car. Hub was quite a guy.

By and large we had a great bunch of people there. Stan Hill was the foreman in the tungsten and Hughie Peters was the shift boss. They were both very good tungsten men.

I was very impressed with the social situation on the mountain when we got there. It was really tremendous that they made sure that no one was put in their place. If you had three children you got a certain-sized house. If you had five you got a bigger one. It didn't matter whether you were a diamond driller or an assistant manager. The size depended on the size of the family, not on the position in the pecking order. It was great. This reveals a lot about the management philosophy at the mine. Everyone got along well at the Emerald too. We had excellent management—well ahead of its time.

At the Emerald I took over from Cliff Rennie as the tungsten geologist. When it closed down, I transferred to the Jersey lead-zinc mine. I ended up spending some time in exploration during the summer until 1960 when we moved to Merritt and the Craigmont copper mine which was then being developed. One interesting feature there was an aerial belt tramline that transported the ore from the pit down to the mill. There I was the open pit geologist and then the design engineer designing all the open pit layouts. It differed from the Emerald in one important aspect—the amount of diamond drilling they did on an ongoing basis. At the Emerald we constantly drilled to delineate the tungsten ore as we mined it. At Craigmont in the open pit that was not necessary, for unlike the Emerald, the ore body was a batholith and had been pretty much defined before we started mining it.

After a couple of years at Craigmont I left mining and went to theological college for three years and then into parish ministry for 14 years. Eventually I got stressed out and came back to the mining business. When I resigned from Placer, Ross Duthie told me that if I ever wanted a job, to let him know. So now in 1979, 17 years later, I gave him a call and said, "Ross, is that job offer still open?" He replied, "Give me a couple of days and I will see what I can do." Within six weeks I was back working for Placer with all the guys I had worked with before—the "Salmo Mafia" they were called. It was a great experience. I ended up being the environmental coordinator for one or two projects. I also wound up in New Guinea doing government liaison work. I took early retirement in 1990 and went back working for Placer on contract for another ten years. They were a great bunch of people and I enjoyed working with them.

Based on my experience, there was a spirit of cooperation rather than competition in the engineering office at Placer. We did a lot about staying abreast of and up to date on the technology. That always impressed me. The lead-zinc mine was one such case with the first use of trackless mining in Canada. Craigmont also served as our training ground for open pit mining, which was a new experience for a bunch of underground people.

The mining business has been good to me and I appreciate that.

GEORGE AND PAT SUTHERLAND

I met Pat and her sister Thelma in 1952 at Salmo, where their father Otto Larsen owned a store and was the local magistrate. The next time I saw Pat was on the Moly Hills Golf Course on the shores of Francois Lake near Endako, BC. Pat meanwhile had married George Sutherland, then the assistant pit superintendent at the mine, whom I met because of a waste removal contract I was in charge of. George began as a surveyor at Canex, but trained as a miner and shift boss. Pat worked in the office. They were there from 1956 to 1960, when they transferred to the Craigmont Mine at Merritt.

I interviewed them at their Penticton home and first spoke with George.

2007 photo of George and Pat Sutherland. They met at Canex where Pat (née Larsen) worked in the office and George surveyed. They now live in Penticton after a long career in mining.

I will be 74 in early 2008, and Pat is a little younger. I was born in Fort Chippewan, Alberta, and took my high school in Edmonton, where I boarded in a private home. Part of the payment for my board was doing chores. I don't know how much it cost my father but suspect it was not a lot. During the summer holidays I worked for Cominco at Pine Point and at their "Con Mine" in Yellowknife. In 1955 my buddy and I came to Salmo and went to work at the HB mine. Late that year I quit and moved to the Emerald, married Pat Larsen, and got a house in the Jersey townsite for $25 a month including utilities.

I started at Canex as a surveyor's helper and worked for Colin Brown while I took some crash courses in surveying. When I was able to survey on my own, Colin went back to engineering duties. Bob Dodds was my assistant, and before long the two of us did all of the surveying in both mines.

In about 1960 I decided that I wanted to get into supervision, so I worked as a miner's helper in various jobs and studied for a Miner's certificate which had become mandatory for all new miners. I followed this up by acquiring a shift boss certificate as well and eventually got a promotion to that rank before moving on to the Craigmont mine at Merritt where I went on to become the open pit superintendent.

I had an embarrassing experience one day. I was asked to go in and fire a "pillar shot" one evening. When I got there, I found two sets of wires coming back to the blasting switch. Two were in one extension cord and two were in the other. The electrician had hooked up the blasting switch, which was 110 volt. His wires were black and white. I took the miner's wires, twisted two blacks together and two whites together, and connected black to black and white to white. I pulled the switch to fire the shot and went home. The next day, a Saturday, the mine captain visited me. He wanted to know "what in the hell" I had done. I told him I connected the wires to the switch, threw the switch, and that was all I did. He said, "It's a real mess in there!" It turned out that the first circuit and the last circuit went, and very little in between. The pillar crumbled and fell over, but there were a whole bunch of holes still loaded with dynamite and caps that had not fired. An investigation ensued and I was asked repeatedly how I had hooked up the wires. Unfortunately, someone had disconnected the wires from the blasting switch, and it was not determined how they had been hooked up. As it turned out after the investigation was over, what should have been done was that one extension cord should have been treated as one single wire, and the other one as another single wire, not black to black and white to white. If you follow a miner's mentality, as I learned later on, that is what a miner does, but an electrician would never do that. Anyway, I wasn't very popular for a while, but it didn't last that long, because within two years I was promoted to blasting foreman. The reason for having two sets of wires was to cut down on the resistance because it was a long blasting line.

Jim and Joan Bristow with Pat Sutherland at the helm of an underground Unimog taxi. Courtesy of Pat and George Sutherland.

Pat then told me a little about her family.

My dad, Otto Larsen, was born in Norway in 1907 and came to Canada when he was six or seven years old. He grew up in Sedgwick near Viking, Alberta, on a farm, but heard from a couple of friends who had been in BC about all the money one could make in the mines. He came out and went to work in the Reno mine up Sheep

Creek. My mother with her two children followed a few months later. She came out by train and lived for a while in Nelson. I was six weeks old when we came.

I lived my first two years at the Reno mine on Sheep Creek where my father was a miner. I was the third of six children—two boys and four girls. The mine was remote. It was on top of the mountain and in winter they went up and down in the aerial tram buckets. In 1938 a younger brother who was born in March died at the mine the following Christmas. Mom and Dad took him down on the tramway and into Salmo for burial. Dad mined for only about four, five years and quit around 1939, but it was long enough for him to get silicosis.

Dad and his brother then bought a logging truck and cut fence posts, hauled them into Salmo and shipped them out on the railway. Next he got into the insurance business. I don't think he had to write exams at that time, and then he broadened out into life insurance as well. Next he joined the farmer's market, and after selling all the feed that came in by rail he unloaded it, and the farmers got it from his store. He later became the local magistrate—I think it was after we were married. Everybody around there liked my dad. He would give you the shirt off his back.

I did all my schooling in Salmo and then took night courses for my bookkeeping. I worked up at Canex for six years in the main office, starting shortly after George and I were married. They then shipped me over to the engineering office, where I was the receptionist and did all the typing for the geologists and engineers. My direct boss was Fitzpatrick, the accountant who was in charge of all the girls.

I met George in Salmo, and we moved to the mine when we got married in December 1955. We lived in three different houses while we were up there. The first one was called a "cracker box." We also got 14 feet of snow our first winter there. We curled, but I don't remember skating. We had no utility bills because they were included in the rent of $25 per month in the first house, but we did pay for our heating oil.

Pat and George Sutherland. Courtesy of Pat and George Sutherland.

Then we moved to the Jersey townsite and paid $30. We stayed there until the beginning of 1962, when we moved to Merritt.

What I remember about being up there was that we enjoyed ourselves. We had many social gatherings and a lot of fun at them. When we had dances, everybody came. It did not matter whether you were hourly paid or staff, we all mixed. The odd fight might break out, but it was not between miners and staff. The biggest challenge was the road—getting up from down below in winter. You could go through several climate changes between the camp and the valley. It could be cloudy below, but sunny at the top where you would look out over the top of the overcast. I can't remember disliking anything about living in the camp. We had a good life there. We played tennis in the summer, skied all winter, and we built a boat while we lived there.

George finished the interview with the following anecdote:

Bob Dodds was in Salmo for the Emerald reunion. He fell and broke his hip when he went to visit his dad's grave. Gwen went back to town to get help, and after the paramedic had him in the ambulance, he said, "That's the first time that I ever carried someone out of the cemetery."

CLEM GROTKOWSKI

Clem's place is hard to miss as you drive by Ymir. His house and yard are crowded with colourful wind chimes, propellers, and other sundry items he creates for the tourist trade. I first met Clem when my wife insisted we interrupt our drive to look at his wares. She settled on a set of wind chimes, little knowing that I would be back one day to interview him about his years at the Emerald. Clem is also a noted huckleberry supplier to his friends in the area. In 2005 he picked 154 ice-cream buckets full of this delectable fruit. Clem worked at the mine from 1956 to 1958 and from 1961 to 1963.

Clem and Bernette Grotkowski in their Ymir home after sixty years of marriage, 2007.

Clem in his carport beneath the propellers and wind chimes that he creates and sells to tourists. His basement is also jam-packed with these items. 2007.

Dennis (Bucket) Waterstreet drilling with a three-boom jumbo. Courtesy of Sultan Minerals Inc.

I was born in Poland December 18, 1924, and came to Canada at age 1½ years, and grew up on a farm in Webster, Alberta.

In 1943 I joined the army and was home on my embarkation leave when the war ended, so I did not get to Europe. I had worked for the Great Northern Railway shortly before joining the army, where I learned to operate a dragline and was paid $0.50 per hour as a qualified operator. My wife and I were married in 1946 and celebrated our sixtieth anniversary recently.

I worked with my father on the farm during the summer and cut pulp wood in the winter. The sawmill owner for whom we worked moved his mill out to BC in 1955 and set up on Porcupine Creek near Ymir. We lost everything when the owner went broke the following year. I next worked on the conveyor system at the Emerald for two weeks in 1961 but could not handle the job of just sitting and monitoring the equipment, so I transferred to ploughing snow and then operating equipment underground for two years.

My next work was on construction of the Tye road in 1957/8. In 1959 we logged in the Prince George area, and I cut fence posts in 1960. Then I returned to run equipment in the Emerald, where I maintained 52 miles of underground haulage roads. Eventually I trained as a miner and operated jacklegs and a three-boom jumbo. My shift entailed drilling, loading and blasting a 10-foot round in a 12-foot drift every day. My cross shift was Jack Walton, who was a great guy to work with.

Unknown man drilling with a jackleg. Courtesy of Sultan Minerals Inc.

No, 12 Caterpillar grader. Courtesy of Sultan Minerals Inc.

My brother Alphonse was seriously injured at the Emerald while he was working on the Giraffe the day his partner was killed when its boom collapsed. He was also one of the men who helped rescue Gil Mosses in 1969 and received medals for bravery. Another brother, Andrew, contracted silicosis after only six years at the mine. He fought a losing battle with the WCB and finally wrote a letter to Premier Bill Bennett and had his claim promptly accepted.

After the Emerald closed, I worked for the BC Department of Highways until it was privatized in 1988. My sister and I are the only surviving members of eight children, even though we are not the youngest of the tribe.

My 15 years at the mine were the best years of my life.

NANCY VERIGIN

There were several Verigin families in the Nelson–Trail area. I did not know either Nick or his wife Nancy, but their names were familiar to me for they were at the mine from 1956 until it closed in 1973. Nancy now lives in Grand Forks, where I visited her in 2007.

I was born in Winlaw, in the Slocan valley, but moved away when I got married. I was quite young, and we lived briefly in Trail where my husband Nick was a truck driver. Later we moved to Vallican, where he worked at the nearby Passmore sawmill. In 1956 we moved to the mine because his friends, Shorty and Harry Kinakin as well as their uncle John Mahonen, said, "Let's quit this job and go to the Emerald mine." Nick was not yet a miner, but he became one at the mine. I was a homemaker there for 17 years, but I also ran the commissary for a while in 1971/2. It seemed that most of the women up there did. They delivered the mail there, so one of our jobs was to sort it. We had a Rec centre beside the commissary. We all enjoyed curling, so it was great to have a curling rink at the mine.

Nick was an underground miner and truck driver all the years we lived at the mine. He died about four years ago, but we had already gone our separate ways by then. We had three sons, all born in the Nelson hospital, who all thought that the mine was the greatest place in which to grow up. They just knew everybody and everyone was friendly. We all seemed

Nancy Verigin lived at the mine with her husband for many years and spent part of the time working in the commissary. 2007 photo.

Nancy's third home at the mine. Courtesy of Nancy Verigin. Courtesy of Nancy Verigin.

Nancy's fifth home (a Panabode) at the mine. Courtesy of Nancy Verigin.

769 Caterpillar truck. Courtesy of Sultan Minerals Inc.

to get along well. Carol Wilson used to babysit my two older sons. My oldest son later became a lifeguard at the pool for one summer

We lived in a couple of "cracker boxes" [*duplexes*] to begin with and finally in a three-bedroom Panabode, which was our fifth and last house.

My sons all live in Calgary, where the oldest one, Jerald, was in carpentry but now owns a company that sells products to convenience stores. He was born in 1953. The middle one, Brian, is in computers and was born in 1957. The youngest, James, drives large trucks. He was born in 1965 and was not planned.

I truly liked living at the mine. It was very, very nice up there, and I loved the snow. To this day, I miss it—seeing all that snow on the trees. It was so beautiful. I never regretted spending seventeen years of my life there.

ROSS MCCONNELL

I visited and interviewed Ross in his home in Salmo, where he has lived since coming down off the hill from the mine. Ross drives truck and school bus for the village and his wife is the high school secretary. He lived at the mine from 1958 until 1973 where his father was an electrician.

Ross's older brother, John, was born on November 7, 1956, the same day as my son Don. He worked at Canex briefly before it closed down, then as a raise miner at the HB mine for four years before going back to school for his engineering degree. He is now the president of Western Keltic Mines in Vancouver.

Here is Ross's story:

I was born in Nelson on May 20, 1958, and lived my first 15 years at the mine. My earliest memories were of having a swimming pool there and the bears. It was just a great place to grow up. About the time I was in Grade 3, I got a little dirt bike that I could ride everywhere.

Ross McConnell, who grew up at the mine, now works for the village of Salmo. 2007.

The bears were no problem, though I guess I had a couple of little scares, for there were some pretty big ones around. When someone shot one, the tow-truck driver would hoist it up and we would all get our pictures taken beside it. I remember some guys would pull out some of its teeth and give them to us. One of our neighbours ran electricity to his garbage can to zap the bears—he had gotten tired of continuously picking up his garbage, but I don't think it helped much.

The swimming pool was great. We'd get lessons in the morning. It would then be open in the afternoon, close for an hour at dinner time, and be open again in the evening for general fun.

I remember playing "kick the can," but I can't remember the rules for it. In winter we used the road down to the curling rink as our toboggan trail.

John McConnell, B.Sc. (Eng., Mining), older brother of Russ. He drove raises at the HB Mine for four years, then got his engineering degree and is now president of Western Keltic Mines in Vancouver. 2007.

We drove into Nelson about every other weekend to visit my uncle who lived there. If it was winter, we first turned off the oil furnace before leaving. When we got back home late in the evening, the house was just freezing. We'd get a piece of lighted toilet paper and drop it down into the furnace to get it going again. The house was so cold—I guess because it was not insulated. We'd jump into bed and lie there just freezing for a while until two or three hours later, then it was starting to warm up again. I don't know whether we turned off the oil to save oil or because we were afraid it might burn the house down while we were away. I remember using "Red Devil" frequently to clean out the chimney before the soot could build up and start a chimney fire.

We got our first TV set when I was about two or three. Not too many people had them then, nor did very many have telephones, and those that did were on a party line. There would be half a dozen people on a single line.

I broke my arm when I was five. I was running home for dinner and just fell and broke it. They had to take me to the Nelson hospital where

Ross with his dog Spot in the early 1960s. Courtesy of Ross McConnell.

Ross fishing with his dad, brother, and another person at Rosebud Lake, early 1960s. Courtesy of Ross McConnell.

Ross as a Boy Scout in the early 1970s. Courtesy of Ross McConnell.

they put it in a "full arm" plaster cast. By the time we got home it must have been very late—ten or eleven. My dad had to rig up a bar over the bed to elevate my arm and to dry the cast.

The teachers I can remember were Mrs. Rose, who I think I had for about four years. I had a Ron Morris for a couple of years. I would have had someone else for a year, but don't remember who it was. The school was fun. It was great. We had our ball teams and we played lots of scrub. As I got older we got into the "Little League," playing for Salmo. I also spent quite a bit of time in the "Boy Scouts." My older

Ross McConnell and brother John, circa 1959. Courtesy of Ross McConnell.

Basketball team in early '70s—back row: Coach Lorenz, Glen Hodge, Ross McConnell, Kenny Crayston, Bruce Gregory, Eddie ??; front row: Tim O'Connell and Mike Pietzche. Courtesy of Ross McConnell.

brother John became a Queen Scout and went to Ottawa on a trip connected with it. When I started school in Salmo we had around 330 students there. In the winter the miners could at times not make it up the hill, but the school bus always seemed able to. Now the Salmo School is down to 160—one half the size. My wife Darlene is the secretary there, and I drive the school bus.

JOHN VERSCHOOR

I met John when I interviewed him at his home on Airport Road near Salmo. His story of arriving in Alberta in the winter and living in a crude shack with no power or running water struck a chord in me, for I grew up in such conditions. He worked in the conveyor system at Canex from 1960 to 1973.

I'm from a family of seven, and my brother and I were thinking of leaving Holland. My mom, who was the pioneer in the family, said, "Let's all go to Canada." It was a good opportunity because the Dutch government was making money available to go there or to Australia. So the whole family left in 1952.

We landed in Lethbridge and went to a little place nearby called Sterling. It was the middle of January with the temperature at −30°F [−34°C] and we moved into a chicken coop. That's exactly what it was, a chicken coop. It had no toilet, no bedrooms, a stove in the corner, no running water, but a barrel outside that we could fill up. What can you do with that when it is minus thirty? It was hard on my parents, and I felt sorry for them. For us young ones it wasn't too bad, we adapted, but for my mom and dad, it was really tough. My dad must have been wondering what in blazes he had gotten into. My youngest sister was only five and learned English in no time. That was good for my mom.

It turned out okay, for we got to move to a place in Taber where we got a nice big house that was warm, and everything had been set up for us including a radio. I can still remember seeing the farmer through the upstairs window come to the door with a bundle of newspapers and a radio.

My dad took a trip back to Holland in the

John Verschoor came to Canada from Holland at age 21 with his brother and parents. He quit farming to come to work at Canex in the crusher and conveyor system. 2007 photo.

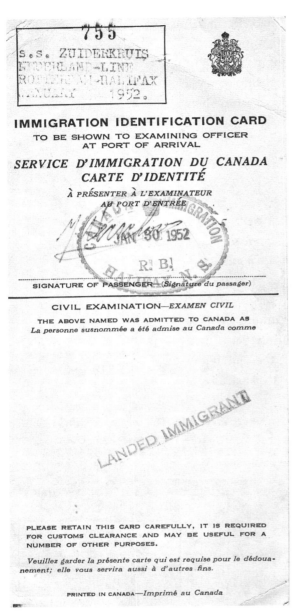

Landed Immigrant document. Courtesy of John Verschoor.

mid-sixties when he was sixty-one. When he got there, it was spring and the grass was a foot high, but here on the prairie there was still snow on the ground. After returning to Canada he did not leave the house for two weeks—he was sick. If he had been on his own, he would have gone right back to Holland and stayed there. I believe my dad must have been about 45 years old when we came to Canada, but things turned out okay in the end.

My mother was the optimist in the family, despite being weak and sick a lot. My dad was a hard worker, a family man who had been a farmhand in Europe. Here he worked for farmers in the sugar beet industry. It was a good experience, but in hindsight, we should have done things differently.

Things changed for me when I came out to Salmo and met my wife. We got married and went back to the prairies for a couple of years. I worked for one farmer where I started at 8:00 o'clock in the morning, then pretty soon it was 7:00 o'clock, then 6:00 o'clock and finally 5:00 a.m. I bitched and complained, but nothing changed. Then one day he left on a holiday and left me to look after a few steers he was getting ready for market. At first you have to go easy on the amount of grain you feed them, or it will be "legs up" if you're not careful. He was gone for two to three weeks, and when he came back, he started bitching and he says, "You're not doing the job right." So he started feeding them and the next morning when we came down, there were two of them lying there, legs up—dead. Meanwhile my father-in-law was saying to us, "Come to BC, there's work out here." So lo and behold, it happened that we came for a visit and I told him, "I'm ready to move." We went back home, packed, phoned a moving company to pick up our belongings, jumped in the car and left. It was the best thing I ever did. We raised four children who are all doing well. It was like a second time immigrating, so to speak.

I wanted to move up to the townsite when I came, but oh no, "We're not hiring anybody." So I drove into Nelson and went to Kootenay Transfer to look for a job. They said, "Yes, you can start on Monday morning."

Passport. Courtesy of John Verschoor.

When I got back home, my wife said, "Canex just phoned. You can start right away. So I had to phone Nelson and tell them I wasn't coming. However, I couldn't move up to the townsite because we had bought this place already for $6,000. It was only a small shack, but it was on an acre of land beside the river. We raised it, put a foundation under it, and eventually incorporated it into the new house which we built around it.

I got a job at the mine and started on the conveyor system for Al Nord and worked with Clare Vayro and Ian Dingwall. I did do some work in the tungsten mill too, but not very much. I worked mostly in the bottom end of the conveyors and became familiar with the long stairway to the Number 4 conveyor adit. One thing about working there was that on sunny summer days we would sit outside the portal during the lunch break. It was tough to get some of the young guys back on the shovel afterwards though. During lunch the guys used to feed the gophers and they soon became very tame. Once in a while when you opened your lunch bucket, you would discover that some prankster had put a gopher inside it. On another occasion in the change room when I came to get my hard hat, I found a pair of miniature wooden shoes hanging from it, one on each side. I never found out who had done that.

We used to have some terrible spills, especially when we came out on Sunday nights or on Monday mornings. After the weekend shutdown we

John Verschoor and Ernie Stenzel at the cone crusher. Courtesy of Ernie Stenzel.

had to start up the belts and open the chute gate from the ore-pass. Then if nothing came, we would blow air or water up to try to bring down the hang-up. Sometimes when it let loose, it just plugged up the whole works and we would have to set up a slusher to scrape out the spillage.

There was plenty of hand mucking as well. It could be dangerous because one worked alone on the night shift much of the time. They'd phone you once in a while to check up on you—big deal. They would ask if you were still there. If you didn't answer . . . anyway, you looked after yourself. It was quite an experience.

Travelling up the hill to the mine was something else, during the winter especially. One time I was coming down and I was going too fast when I hit a hairpin turn. I realized I was going too fast, but there was no way I could slow down. So just as I hit the turn, I hit the gas pedal and cranked the steering wheel at the same time. The car went sideways, bounced off the snow bank and away I went. It didn't even dent the side of the car. I was sure lucky.

The one thing that really shocked me was the amount of drinking that went on. I don't spit in a good drink, but this was something else again. They used to throw parties down at Rosebud Lake. Alcohol! You wouldn't believe it! Or the New Year's party up on top? I just couldn't believe the amount of alcohol consumed.

Ian liked to drink, so once when he and his wife were going back to camp, he spun out right at the bottom of the hill. His wife got out to push the car and when Ian got going he never stopped. He left his wife standing on the road at the bottom of the hill. I guess he forgot her.

When the mine shut down I went to the HB and got hired on. Later I was working for Cominco in Trail, where they said, "We know you can do the work, but you have no qualifications to be a millwright." So I worked as an "oiler" for a while. I tried unsuccessfully to write the exams, but I had not been in school for many years. The guy who was in charge of the apprenticing told me they had upgrading courses in Nelson that would take only six weeks to complete. He said that if I passed that they would give me a millwright's job.

Before starting my classes, I said to my boss, "I gotta go to school for about six weeks. Will Cominco pay my wages while I'm doing that?" He did not want to discuss it, so I continued, "What if I pass and get my qualifications. What about then?" He said, "We don't want to hear about it." So I assumed it was company policy. I took a month's holidays and a two-week leave and went to school. I got my ticket and got my millwright's job.

Meanwhile Eddie Bengert had been doing the same thing. He too had gone to school and gotten his millwright's ticket. So I asked him if Cominco had paid him while he was in school. He said, "Oh yes. I got a cheque every two weeks while I was in school.

I then confronted the foreman with this and asked why they would do it for one and not for the other. His first concern was whether I had gone to the union with the issue. Later that same afternoon he came back to me and said, "Your money is in the bank."

One event I never told anyone about was opening and dumping a barrel of cyanide in the Tungsten mill without wearing a respirator. Did I ever get sick! I thought I had come down with the flu. There were chemicals all over the place then and I sometimes wonder how I survived.

I spent the rest of my working days at Cominco in Trail and enjoyed

Governor Generals Award to Debby's mother-in-law, Jean Swedberg. Courtesy of Debby Swedberg, née Shelrud.

working for them. Coming to Canada, though hard for my dad, was a good decision. I have done well, and my children, who are all professionals, are doing even better.

THE SHELRUD FAMILY

Carl Shelrud was an important figure in the Gil Mosses rescue. When I contacted his widow by phone in early December 2006, she agreed to an interview, but when I later tried to arrange one for early January, I was unable to reach her. I finally learned that she had passed away around Christmas time. I was able to track down two of the Shelrud daughters, Debby and Gail, whose interviews follow.

Debby Swedberg has the distinction of having a father as well as a mother-in-law who were awarded medals for bravery. Debby's husband's mother, Jean Swedberg, was working at the desk of the Valnicola Hotel in Merritt when a fire broke out. She and Mr. Sykes, a principal of the hotel, were able to alert all the guests except one, who along with Jean perished in the blaze. Her family received posthumous awards from both the WCB and the Governor General of Canada.

Shelrud wedding. Courtesy of Debby Swedberg, née Shelrud.

Family sitting on the stairs. Courtesy of Debby Swedberg, née Shelrud.

DEBBY SWEDBERG NÉE SHELRUD

We moved to the Emerald in 1960 and lived there for 10 years. I went to school there from grades 1 to 6. I think we had between 30 and 35 kids in the whole school. I took grades 7 to 9 in Salmo. We moved to Britannia Beach in 1969 and from there to Merrit two years later, where Dad worked at Craigmont.

The teachers I remember from the mine were A. Kowalyshyn and M. J. Haman, both principals, as well as Anne L. Postlethwaite and Glenda J. Ryall.

We spent much of our time as youngsters hunting frogs and snakes at the Tungsten tailings pond. We rode rafts on the pond, but of course there were no fish in it.

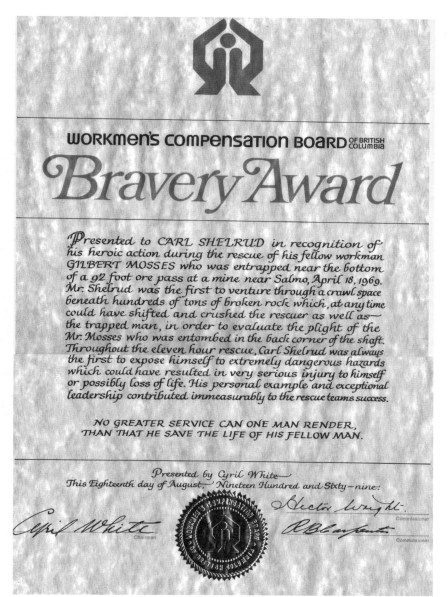

WCB Award to Debby's father. Courtesy of Debby Swedberg, née Shelrud.

There were about 100 families at the mine. It was a really fun place to grow up when we were kids. We could do anything—fishing and hiking and all that stuff.

One spring we had a fire in the school, so they shipped us to the community hall. It was over close to the main camp. They had an amazing library there. I'd love to get my hands on some of those books now.

I remember one year the roads were so bad the bus got stuck coming up the hill. We had just the one bus driver and he used to get harassed—that poor man. His wife was a teacher in the school in Salmo.

GAIL BENTLEY NÉE SHELRUD

I knew Sheila McLean's mother really well. She used to serve tea and cookies to all the neighbourhood girls. She was good at getting Terry Owen and I together after he moved down to the south camp, off the hill. She took us to Trail for an outing once. I was the youngest girl of five children, and she gave me some much needed affection. I do have a younger brother.

When I was in grade 6 at Canex there were four kids in grade 4, five in grade 5, and six in grade 6. While I was in grade 3, we had a hot-tempered teacher who was French. One of the Craig girls was so painfully shy that she could hardly talk to anyone or make eye contact—she was that shy. One day when she put her hand up, the teacher demanded, "What do you want? What do you want?" It was obvious to us all that she needed to go to the bathroom, but the teacher just kept demanding an answer, until finally there was a puddle on the floor. I guess the teacher was trying to get her to learn to speak out. On another occasion, when she caught the Summers' girls smoking, she grabbed them by the hair and hauled them up the hill. We did sometimes get some real winners, but the place was Paradise. It was just a wonderful, a wonderful place for kids to grow up.

In 1963 or '64 we had a fire at the school, and we went to school at the Community Centre (the "Hall") instead. They bussed us there in the "panel truck," which was also used to haul mail and supplies to the mine.

Spring was the best time. There was a creek with a large clay deposit nearby, where we could go and build small dams with it. We'd also climb up on a hill we called "Baldie" where there were some large rocks that got very warm in the sun while there was still snow on the ground. We loved that, for my sister and I would take our shoes off, then walk barefoot in the snow, competing to see who could do it the longest before having to climb on to the warm rocks to get the feeling back into our feet.

The flowers were just unbelievable up there. There was one spot down by the pool where the lady's slippers grew. We took a footpath that went from the circle, down through the ravine, and came out at the back of the school. There in a wooded area was the only place we could find lady's slippers. I hope we didn't kill them off, for we picked so many. In the spring we'd get glacier lilies, tiger lilies, Indian paintbrush, violets, and some tiny little pink and white flowers that I have never been able to find in any book. There was a vine with a very pretty blue flower—it took me years before I found out that it was wild alpine clematis. We used to explore a lot and sometimes get into trouble by going into restricted areas outside of our "boundaries."

Canex was a really family-oriented company. We had Santa Claus come around every Christmas. One year at Halloween someone lent us a Donald Duck costume, and because it happened to be my size, I got to wear it for the costume parade.

The highlight of each summer was the company picnic at Sullivan Lake. I remember winning all the races there, but I'm sure I couldn't have won them all, but that is what it seemed like to me. We were all athletic—how could we not be, growing up on a mountainside? We wandered all over the hillside in the summer, picking wild strawberries and trapping gophers. Murray Crayston's son Donald taught us all how to skin a gopher.

We used to go down to the dam on Lost Creek, where we sometimes swam (it was the townsite water supply). Once when we were on the dam throwing sticks and rocks into the sluiceway, Dad started yelling, "Get to the truck! Get to the truck!" We looked up and saw a wall of water coming down the creek. We hightailed it across the dam and down to the truck. When we looked back up, the dam was engulfed in water. I think a log jam somewhere back up the creek must have let go.

We had to learn to look after our garbage because of the bears. Dad shot one off our back porch one day that had been trying to get into the garbage there. When they hauled it away it almost filled a dump truck, it was so huge. The last house in which we lived was in the centre of the townsite where we weren't bothered much by bears. In later years they began to bring in "barrels" for live-trapping the problem bears.

Murray Crayston, 1967. Courtesy of John Bishop.

Murray Crayston measuring something. Courtesy of Sultan Minerals Inc.

Lost Creek dam—good fishing too. Courtesy of Cliff Rennie.

The skating rink was just a piece of flat ground that they flooded each winter, but it was enclosed with a low board fence. We couldn't afford skates, so I don't remember skating on it. They had to keep a hose running to flood it occasionally, and the spray from that hose would cover the trees like freezing rain and the slope down the hill would ice up. It was more fun playing on it than on the rink. I never really learned to skate.

I think we paid $0.50 per hour to rent a horse from Feeney. That's where I learned to ride.

Mom was president of the curling club for a couple of years. She also had a job cleaning the office, sold Avon, and did stuff like that.

Dad was always active in mine rescue and first aid. He trained the mine rescue teams and would take the team down to practice by the tailings pond while we'd chase around catching frogs and snakes. We children also used to ride around on rafts we'd built out of loose boards we'd found there. Once when we flipped over some boards there, it was like uncovering a moving blanket of baby frogs that were immediately hopping all over the place. I have never again seen anything like it—hundreds of them hiding from the sun and the birds.

I don't remember the Panabode logs shrinking, but some knots would shrink and fall out and create peepholes. They were really neat houses—dirt basements, but we didn't know any better. When we moved to Britannia, we moved into a house with an ensuite, a walk-in basement, and my brothers had a room down there because it was a real basement. After those shacks at Canex we thought that we were in a mansion.

When I was twelve, we moved to Britannia Beach and later to Merritt, where we were a little better off financially and my parents bought a pair of skates for me. My older sisters resented me for this because they'd had to get out and earn their own money for skates.

My dad never got the mining out of his system. He tried some independent ventures and prospected until his health failed. He passed away in 1998.

ED LAWRENCE

I met Ed Lawrence at the Salmo Reunion of 2006, and during a conversation we had a couple of weeks later he suggested that I write a history of the Emerald. When I asked if his company would sponsor the project, he directed me to Art Troup, the president of Sultan Minerals Inc., the company that now owns the property. Sultan Minerals Inc. has indeed sponsored this project generously, and Ed Lawrence has generously shared his and the mine's stories and photographs with me as well. He worked for Canex in several positions from 1962 to 1974, including that as mine manager. He was the manager when the Gil Mosses rescue took place, and the Workmen's Compensation Board awarded him a silver metal for bravery for his part in the operation.

Ed lives at Westbank—whenever he finds time to be at home, for, as this is being written, he is in charge of exploration for Sultan Minerals Inc. at the old Canex property at Salmo. He has a son, Michael, who works in computer support services at Westbank, and a daughter, Lisa, who lives in Peachland. Ed is yet another example of a man working well past normal retirement age.

I was born in Vancouver in 1936. My father worked for Goodyear Industrial Products, and my mother had a well-paying job as well. I attended schools in Terrace and Nelson and graduated from UBC as a mining engineer in 1959. My first jobs were with Kennco and Rio Tinto before coming to Canex on January 6, 1962. I will not soon forget the Dawson Creek temperature dropping to –54°F (–48°C) one winter when I was there with Kennco.

I started with Rio Tinto on the west coast of Vancouver Island in 1960, then in the Highland Valley, where I worked for a Lyall Gately. I left Rio Tinto and went to work for Kennco and was doing some work in the Bob Quinn Lake area the following year.

I was doing a traverse on my own and was climbing a steep cliff face where the footing got difficult. It got to the point that I could not retrace

Ed Lawrence, B.Sc. (Eng., Mining), the last mine manager at Canex, with Stan Endersby. Ed is now in charge of exploration at the Emerald for Sultan Minerals Inc. 2006.

Ed Lawrence, mine manager; Bert Lundeberg, shift boss; Jack Robinson, mine superintendent.; unknown; Utica mine manager; Gunnar Adolphson, mine captain; Ron Stard, surveyor. Mid-1960s. Courtesy of John Bishop.

my steps, but had to continue up. The terrain finally became gentler, when to my horror I saw a Grizzly sow with two cubs not more than fifty feet away. I slowly backed down the hill as far as I could and waited, but nothing happened. After a thirty-minute wait I inched back up to the brow of the hill, but the bear was gone.

Fog rolled in about the time the helicopter was due back, so I resigned myself to a night on the mountain. I continued the traverse and found an old cabin that had been sunk into the hillside. It had no windows, so I lit a candle to help explore its insides. I found an old steel bed next to several cases of antique dynamite that was obviously unsafe to move. I had no intention of sleeping next to that, so with great care I moved the bed outside and built a fire.

The night was cold and I had no sleeping bag; just a jacket. It was not until after sun-up that I finally slept fitfully for a short time before being awakened by the sound of the helicopter passing. I came wide awake in a panic, thinking that the chopper was gone, but it soon returned and picked up a much relieved passenger.

I got married and worked for Kennco in the Alice Arm area that year, where my wife joined me. It was getting into December, my wife was far along in her first pregnancy, and there was no guarantee of work for the winter, so I contacted Doug Little at Placer and got an interview. It was not long before I got a telegram from him with an offer to start work the first week in January at $425 per month. Although this represented a cut in pay, it was a welcome turn of events for it was a steady job, and my son was born shortly after we moved to the Emerald.

My career at Canex included positions of chief engineer and assistant mine manager.

I served as the last mine manager from 1968 until the mine closed in September 1973, but we stayed on until May 1974 to oversee the final disposition of the mine's assets.

One thing I liked at Canex was that we had innovative people. The mine became known as the Canex School of Mines because of the many

Sue Lawrence; Owen and Joan Bradley. Mid-1960s. Courtesy of John Bishop.

Dennis (Bucket) Waterstreet operating a three-drill jumbo. Mid-1960s. Many people did not know that the New York Rangers hockey team had signed Dennis to a "C" contract, but Dennis must have liked the security and the size of the Emerald paycheques better. He worked there for 14 years. Courtesy of John Bishop.

people we trained there for other operations. That included surveyors, geologists, engineers, and miners. One example was in 1972, when I hired Ron Stard from Salmo who had just graduated from Grade 12. Ron started as a surveyor's helper. He shortly took over as surveyor, then as "mine engineer." He left when the mine closed in 1973 and later became the manager at a large open pit coal mine.

I remember Chuck S. Pillar, VP of Operations, who was known as the "Arizona Rattlesnake." He was in his sixties, childless, and viewed the managers as his sons. Though reputedly hardnosed, he also had a soft streak.

We did about 70% of our output with jackleg-mounted drills. The rest was done with a three-drill jumbo, which Waterstreet, Fleming, and Kinakin operated most of the time I was there.

The Invincible adit was a 10,000-foot decline. The tunnel was 17 feet wide by 19 feet high and used a 5-foot diameter plastic vent pipe for ventilation.

The conveyor system at the mine was unique for the era. At the time, it was the longest conveyor system in the world.

Things were not always dull. In 1969 the Salmo River flooded and gave us a good scare. We had to quickly rip-rap the tailings pond to prevent the river from washing part of it away.

I was the mine manager when Gil Mosses fell into an ore-pass and was trapped. Together with the assistance of shift boss Carl Shelrud I supervised the dramatic twelve-hour rescue operation. *[Details of this event are found in Part IV of this book.]*

My best times as a family were my years at Canex. The least enjoyable was dealing with Legal and Accounting in the Vancouver head office. However, Canex was largely an independent operation, and Head Office pretty much let us run the show from the mine.

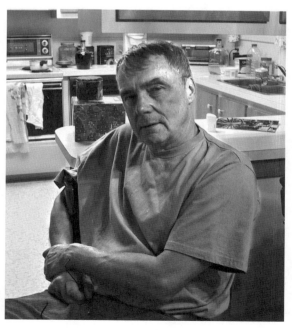

John Bishop in 2007. John was not only an accomplished surveyor, but a great photographer as well. A large number of his superb pictures can be found throughout this book.

JOHN BISHOP

I first met John when I interviewed him in early 2007. He provided me with thirty-one excellent photos that he had taken. He told me he paid $350 for his camera in 1960—a lot of money back then. Based on the quality of his pictures, he obviously learned something about photography as well.

Salmo River in flood, 1969. Left to right; Albert Fontaine, Bob Adolphson, Bruce Martin, Alfie Ebert, Dennis Rogers, Roger Martin. The bridge was swept away 30 seconds after the men cleared it. Courtesy of John Bishop.

Albert Fontaine with his custom VW Beetle. Late 1960s. Courtesy of John Bishop.

John has been doing exploration work at the Emerald for Sultan since early 2007.

I grew up in St. Margaret's Bay south of Halifax, and ten miles from Peggy's Cove. My great grandfather was lighthouse keeper at Peggy's Cove. I don't have a university degree. I went to Nova Scotia Land Survey Institute. To finish the course, I would have had to work under a licensed Provincial Land Surveyor (PLS) for a year and then write another exam, but I came out here and tried the underground work for a year. That was it. I did get a pretty good grounding though.

I came to Salmo in June 1964 and moved in with Scottie Thompson, an assayer. The guys said, "You're not moving in with Scottie?" (He was not noted for cleanliness.) I stayed there for a couple of months until they found a duplex for me where I lived next door to Werner and Heidi Trachsell. *[The staff house that the men stayed in earlier, had burned in 1953 and there were no bunkhouses left either.]* After a year I moved down to one next to Harold Wainwright. He was the head warehouseman. Colin Brown became the manager just after I arrived, a job he held for a couple of years before he moved on.

I left Canex and went to Jedway for a year, and then back to Nova Scotia, where I surveyed for a while. While I was at Jedway Iron Ore, we drove one raise up 400 feet to meet a sub-drift, which we hit the drift dead on. However, one day Canex management called me and asked, "Would you be interested in being a geological technician?" I asked, "What's that?" Anyway, I came back to Canex in January of 1967.

At the mine we were still in the lead-zinc and expected to stay in it for a while, but when the NDP formed the government and threatened to change the mining climate with a super royalty, the HB, Reeves McDonald and Canex all abandoned the low-grade zinc, because they could not afford to mine it any longer. The tungsten kept us going for a little while, but I was gone in 1972, before the shut-down. I went to work logging for Ole Jensen of Salmo. He had earlier told me that anytime I needed a job, to come see him.

I married Donna, Bob Rotter's daughter. Donna and I are divorced now, but we have three beautiful daughters: Jeannie, Patty and Linda.

Ken Henderson, Donna's uncle was an accountant at Canex, way back. He was married to Bob's sister Fran. Both Bob and Fran are gone, but Ken is still with us.

I'd be remiss if I didn't mention my father-in-law, Bob. He had a lot of patience with this Nova Scotian, who didn't know dick about mining, logging or carpentry. (Some people might say that I still don't). Mind you, there were times when I showed a bit of patience myself.

My time at the mine was from June '64 to November '65 when I was a surveyor when I got into producing mining plans. We would have a little ore body, and I would draw the manway raise and I would draw up the ore-pass and the plans on how to mine it. I found that interesting. My time as a geological tech lasted from January 1967 to May 1972.

Owen Bradley and Gunnar Adolphson, mid-1960s. Courtesy of John Bishop.

Now I am up there yet again with Sultan Minerals. I started off in the office going over a lot of the old plans. I was one of the few people still around who had worked there and was familiar with the old plans and layouts. We had done most of those old plans at 20 feet to the inch, so we are changing everything to "50 scale." A lot of the plans we can send to Kelowna to get converted, but some we have to do by hand. I am now doing a fair amount of surface mapping. I am also doing some surveying and laying out the diamond drill holes and logging and plotting them. We just had an outfit from Calgary who came in and did a laser scan of the mine in 3D.

They're paying me well and for an old guy, it's great, but gee, I'm doing what I did before—that's why I quit the first time and now I'm doing it again. Actually we're finding stuff that we did not know was there. We're getting into a new area. Finding ore on the surface and mapping that. We see good zinc and then we drill down and we see moly, and the stock market is coming along—and its all kind of interesting. So that's a help to me.

The only close call I ever had was one day when Owen Bradley and I were in a stope. We stepped out to light a smoke. I don't know why we did that, but while we were out there was a big fall of rock where we had been. I had completely forgotten it, but Owen reminded me of it.

To make a mine out of the place again, we just

Mary and Harold Copley with Barry Thompson (John Bishop's cousin) in back row with children Nancy, Betty, and Bobby in front row. Mid-1960s. Courtesy of John Bishop.

JEWEL OF THE KOOTENAYS / 227

General and engineering offices, mid-1960s. Courtesy of John Bishop.

Old Office; with Gunnar Adolphson, Bruce McNeil, and Bill Thompson. Mid-1960s. Courtesy of John Bishop.

need to prove up a million and a half tons of zinc, and we've got over half of that now. If we can do that, it will be viable and we can throw in a mill. With the new technology that has come along, I'm sure we can do it cheaper and more simply than before.

CLARE VAYRO

I met Clare in his home across from the Salmo golf Course on Airport Road. He worked on the conveyor system at the mine from 1964 until the mine closed in 1973.

Clare Vayro in 2007.

I came here from Taber on a holiday. When I went to the mine, Al Nord offered me a job, but I told him to hold it open for a week so that I could finish my holiday and go back home and quit first. When I returned the following Monday, they hired 27 people that morning of whom 26 were from the HB which had just closed down. I was scared—almost shitting myself, thinking, I have quit my job back in Alberta and here are all these experienced guys ahead of me. I will never get a job now. Just the same, Nord kept his word and hired me.

The mine sent me to Dr. Carpenter for a medical. When the doctor examined me, he immediately said, "You smoke too much and you drink too much." I did indeed smoke a lot, but I didn't drink at all.

You had to work at the mine for a month before they would give you a house. After working for six months, I bought a little camping trailer and brought it up there. The company actually built a spot for it. They installed power, water, and a septic tank for it, and I lived in it that winter. In the spring after I had been at the mine for only six months, they let me move into a house that had been empty from the day I first went to work there.

Howard Lee and I worked together in the crusher. There were a lot of rollers and belts to maintain and we often got muck spills that had to be cleaned up by hand. When we left on Friday nights, everything was always fine and dandy, but you could get a hell of a surprise when you came back on Monday. The crusher would handle about 100 tons of rock per hour. If we could get the muck through the chutes, the jaw crusher could handle it. Whenever a rock was too large, we would hoist it out of the chute, drill a hole or two in it and then drive wedges into the holes and the rock would split like a damn.

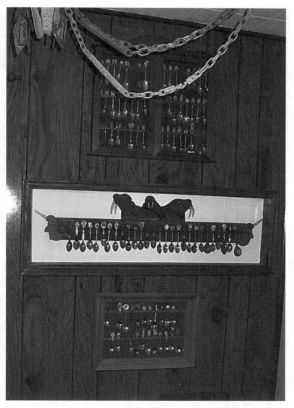

Vayro's spoon collections, 2007.

There were times when we worked seven days a week, but at the lead-zinc it was pretty much a five-day-week operation. When we got into the tungsten, it became a seven-day operation, and we worked on a six-and-two schedule (six days on and two days off) and changed shift every two weeks. In most cases when you were on the conveyors, there was always muck to clean up with shovels. Conveyors number 3 and 4 were actually pretty good, but not numbers 5 and 6. That's where the water was, and that's where we got the blowouts. There was many a time that we would shut down the system and the entire crew would go down to 5 or 6 and pray it would be 6 because it was a hell of a walk up the hill to 5. As I remember it, there were 685 steps up the stairway to get to it. There was no road, so you couldn't drive to it. You had to do it on foot. There was a tram for servicing the conveyor and in all honesty, I don't know how they built it—you couldn't put a helicopter pad there for it was right on the face of the mountain. There was just a walkway beside a set of rails and a little flatcar to transport materials and equipment to the site.

I rolled my old jeep one day. Everything seemed to be happening in slow motion. I stayed in the vehicle, but it threw my father-in-law

Looking up the skipway. Courtesy of Sultan Minerals Inc.

Timbering in bad ground. Courtesy of Sultan Minerals Inc.

fifteen or twenty feet. Of the two kids in the back, one was still sitting there like he had been before it rolled, and the other kid was lying on top of him. My father-in-law had a sore back that the swimming pool sometimes alleviated, but none of us were injured.

When we shut something down for maintenance, we locked out the electric motors with combination padlocks, and unless you shared the number, no one could start up a motor you had locked out. That way there was no danger of a machine starting up if you were working on it. I still have my old locks out in the garage, but I can't remember the numbers anymore, so they aren't much use to me. I've still got my old miner's belt too.

In 1973 I was working for the auctioneer, Ritchie Bros., when we tore out the crushing plant. They sold the conveyors "in place" just as they were—still set up in the tunnels and on surface. Ritchie Bros.' supervisor told us that it was one of their larger sales of that time. I believe that they actually bought some of the stuff themselves and sold it off piecemeal in other auctions over the next several months.

I haven't been to the mine for a long time—I think you now need permission to enter. There are about four families who live in the old townsite. The old houses were all moved out, and there's not much left there. They closed off the old portals, but as fast as they did that, someone always reopened them. The scavengers have been there, but since the mine was trackless, there were no rails to steal. I'm told there is an old dozer in there somewhere. It was actually a fairly new one, but a section of the roof came down and they would not let anyone go in to get it. An eccentric millionaire bought Vectra Village (the old mill townsite) and rents out the houses there. He reputedly has some houses in

Al Nord cleaning up a "blow-out" spill with a slusher hoist. Courtesy of Al Nord.

Salmo as well that he rents out, yet he spends his spare time picking up bottles along the road.

It would be ironic if I went back to work at the Emerald now, for that was my first job in BC and it would really be something if it was also my last one. I am not sure I can afford to retire at 65. We had it all figured out—we replaced our major appliances, the washer and dryer, etc., but can a guy live on his pension? We have a camping trailer and want to do a lot of travelling, but with the price of fuel and campsites, I don't know. When we went to Canmore recently, we got the last two (unserviced) spots in a 250-site park and paid $38 for each one. We were crammed so close together that our awning was almost touching the slide-out on the next trailer.

Today you can't go wrong getting an education. A high-school diploma is now worth nothing, but it was all one needed when I was young.

HOWARD LEE

I met Howard at either Tim Horton's or the A & W at Fruitvale in the winter of 2006/7. He had just retired.

Howard worked at the Canex crusher for four years, from 1964 to 1968. He found it was a boring job when things were running smoothly. It was just a

Howard Lee worked at the crusher and conveyor, then took a temporary job with the BC Liquor Control Board that lasted 35 years. 2007.

matter of pushing either the "stop" button or the "go" button. After a short time on the job he could tell how everything was going just by listening to the pitch of the electric motors. He notes that he read between one and two paperback novels almost every day.

Howard told me he had been on the job for only three to four months, when he was promoted to operator status because of someone retiring or getting sick. He was 22 years of age, but the mine frequently sent helpers who were 20 to 30 years his senior. He remembers one occasion when the electrical work was being downsized, Adam Wagner was sent as his helper. Adam was about 50 and had been on the electrical gang for quite some time. Howard says that he occasionally took his turn on the shovel to give the older man a break.

The crusher was situated on the 3,800-foot level and fed the conveyors that took the crushed ore to the mill far below. Running smoothly was not a continuous affair. Sometimes the muck would hang up in an ore-pass and the crew would try blasting it down from below, using high-pressure air and water. At other times the ore contained so much water that it would spill over the belts and fill the conveyor tunnel.

Howard recalls one scary incident from about 1966, when a temporary weekend supervisor authorized the dumping of saturated muck into the ore bin. Flow into the crusher was controlled with the use of ten very heavy chains, the individual links of which measured between 16 and 18 inches in length and were built of bars that were 2 to 2.5 inches thick. When the wet muck hit these chains, it did not even begin to slow down but spilled over both the jaw and the cone crushers. Howard yelled at his helper to run and turned off the motors as he too ran. The muck filled the chamber so quickly that he was not at all sure that he would not be buried alive. Afterwards he realized that it could not have enveloped him at his station, but his helper had been in mortal danger.

Another episode he remembers was the winter of 1967/68 when the temperature dipped to minus 20°F (almost −30°C). Everything froze up, including the water pipes. The crew had to cut the bolts off the victaulic couplings and bring the pipes into the crusher plant to thaw them out.

Life at the crusher was not without its lighter moments. There was a security guard who tended to stick his nose into what did not concern him. One day when the crusher was stopped for maintenance work, he came over and demanded to know what was going on and why they were shut

Rolling up belt on No. 6 conveyor. Courtesy of Al Nord.

down. They told him that a big rock was stuck in the jaw crusher and sent him up to the shop to fetch a "sky hook" with which to lift it out. For a man of his age he was remarkably gullible and proceeded on his fool's errand. At the shop, they played along with the gag and sent him to the office, where they too played along and sent him looking for it elsewhere. It seems it took the man some time to catch on. On another occasion Howard and his crew sent a new helper to fetch a can of "compression" for an engine that was not operating smoothly. Whenever the crews were doubled up while changing liners in the crusher, they often played good-natured little pranks on each other. One man snuck up behind Howard and banged him on his hard hat with a wrench. Howard had the last laugh though, when the fellow dozed off during the lunch break. Howard hit his hat with a hammer hard enough to break it!

Though Howard did not work with the miners, he knew a lot of them and said, "Pretty much every one I knew was a little bit crazy." Of course that quote applied to others as well as to the miners. His own father had been a miner too and a very good drinking buddy of my friend Rae Thomas in Ymir.

In 1968 Howard quit the mine to go back to finish his high school. After graduating he took a temporary job with the BC Liquor Control Board. He intended to attend university and go on to a bigger and brighter future. However, when I interviewed him, he had just retired after 37 years with the Board.

MERLE MAHAR

I met Merle and his wife Gloria in their home in Salmo when I interviewed Merle for this book. His is one example of how one's life can be changed in one split second. But it also attests to the fact that a calamity can be overcome with courage, patience, and the support of loved ones.

I was born December 20, 1941, at Truro, Nova Scotia. I came to Ymir with a bunch of other guys who had all worked in the barite mine three to four miles from Walton, where I had helped load ships as well as doing other jobs.

I worked for Connors at the Emerald from 1964 until getting hurt on May 17, 1967, when I lost one eye and part of my face. My forehead had eggshell fractures that took years to heal. I had plastic surgery five times over the next four years and later had pins through my cheek and nose to keep the eye socket in place. The doctors have done a superb job in reconstructing my head, but I suspect that few of us would want to go through the agony that I have suffered. My "partial disability" pension started at only $54.45 and is presently $330 per month.

Gloria and Merle Mahar, 2007. Merle was a diamond driller who worked for Connors and Canex before being disabled in a serious accident.

I married Gloria née Burgess a few months after the accident on a postponed wedding date, and we have a son and two daughters.

The WCB sent me to Accounting School, after which they found a job for me in payroll at Remac, but I found the strain on the good eye was too much. I tried working in the warehouse, but made so little money that I left and went into heavy construction after getting into Local 168 of the Tunnel and Rock Workers Union.

I worked on numerous jobs in mining, construction, and logging in both the West and East Kootenays. This included the Seven Mile Dam near Trail and the CPR tunnel in the Rogers Pass east of Revelstoke. The CPR tunnel was the worst job I ever had because of the bad air from the blasting. One thing I liked about the Emerald was that it was dry and had good working conditions.

JAMES (DALE) BURGESS

Dale Burgess, among several others, came from Nova Scotia in the sixties, but I didn't know him or any of the other Nova Scotians because I left the area in the spring of 1962. He was one of the very first people I interviewed for this book. I saw him at his home across from the golf course on Airport Road near Salmo in late 2006. I have never been a person to keep old material, but in writing this book I have learned to appreciate those who do. I was able to scan a number of items in Dale's possession, including the Carnegie Medal. Dale was at the Emerald from 1964 to 1973.

Dale Burgess and Kay Chisan, 2007.

I was born on January 14, 1945, at Walton, NS (between Yarmouth and Halifax). I came to Salmo with Billy Gould in 1964 when I was 19 years old, and am not related to any of the other Burgesses around here. The only reason we stopped in Ymir was because Noble Gould lived there and he was Billy's dad's brother. We found that there was a lot of work and a lot of women here. There were fourteen men from my home town who came, and many of them are still here.

I first worked on the concrete abutments across from Rotter's sawmill on the Creston highway, which they had finished a little earlier. I also worked for Noble Gould a little while in the bush. Then I got on at Canex about four or five months later and worked there until they closed her down. I learned to run equipment here and it stayed with me, but I drilled and blasted too. I liked to do this on my own, because nobody bothered you in those days. I went to HB for one year after Canex closed but didn't much like it. I have never drawn unemployment and I have enjoyed living here.

I used to run both grader and dozer underground. We hauled ore out with DW10 tractor and trailer and also used 14-ton Eucs. They were the modern stuff. We used the old Dart trucks for nipping vehicles, and I even took out the "honey boxes" and stuff like that on them. They made a man out of me pretty fast. If you couldn't mine you wouldn't be here long. One didn't get much training on anything. On the drilling I would watch what my partner was doing, and it didn't take too long before I knew what to do.

Dale finds "Smallwood" in Salmo. Late 1960s. Courtesy of Dale Burgess.

I worked for a while in the highball drift with Bucket Waterstreet. When he blasted, I would wait for the smoke to thin out, and then down I would go and muck it out as fast as I could and help him get set up again.

The diesel smoke in the mine was absolutely terrible, but it doesn't seem to have harmed me for I'm okay now, but it sure bothered me then. They didn't maintain the equipment too well. You'd start them up, and the water in the scrubber would soon leak out, and a lot

Dennis (Bucket) Waterstreet operating three-boom jumbo. Courtesy of Phil Steele.

of the equipment had no brakes much of the time. I also ran Scooptram—a machine that would quickly heat the drift with that big V8 diesel Kohring engine. The machine was 30 feet long with a 5½-yard bucket on it. It was the first one around here and Phil Steele from down the street and I ran that bastard for a long time. It was a powerful machine, like very powerful! We installed tire chains all way around on it, trying to keep tires on it. It was brand spanking new and it was pretty nice to get on it and run it. We operated on three shifts and sometimes seven days a week.

Down in the Invincible we were on contract, using a Cat 988 loader and 769 Cat trucks. We used them to build new road all the way down to the office. That was all done on the side on the weekends. We got paid so much per trip. Those trucks were big, but so were the 30-by-30-foot drifts we drove them in.

A Caterpillar 988—five-cubic-yard front end loader. Courtesy of Phil Steele.

Directly below the Dodger is where the Invincible goes down underneath. We often helped the miners to load their rounds so that they could get it shot and give us more muck to haul. We were driving the tunnel in solid waste, but then we made a switchback and went down into the tungsten ore body. It didn't last long. We may have mined down there a year and a half to two years and that was it. So I doubt it ever paid for itself. It was also where we hit that big lake underground and flooded the drift for a week.

Bucket was drilling when he hit the water, the pressure so great that it pushed his drill right back. He pulled the steel out, and the water just poured in, and that part of the mine was flooded for over a week. You should have seen the crystals we found. We brought buckets of those crystals to Skyler Peters, who embedded them in the handles of his knives. *[Skyler was famous in the West Kootenays for his knives.]*

I lost a best friend (Wayne Johannessen) who was killed when the Giraffe he was on collapsed. I went to the hospital with him that night and was the last one he talked to. He was scaling and Alphonse was at the controls when a pin that held the two booms together broke. He fell out over the cage onto the rock below. Alphonse was hurt too. My father-in-law, Jim Gray, had run it for years. He had run it on day shift and had been complaining about that pin, but nothing was done about it.

Giraffe—reaching the high places. Courtesy of the Nelson Museum Archives.

We used that Giraffe for everything, not just scaling. We drilled rock bolts and hung wire mesh from it and used it for guniting, (blowing concrete). We also had a hundred-foot Giraffe in there. It was on a "White" truck and the cage went up 95 feet. The only way we got her into that Jersey was by letting the air out of the tires to get by one timber set in the "A" zone. There were places in that "A" zone where we built up ramps and still couldn't reach the back with it. That was what it was purchased for, that high-grade "A" zone. I'll tell you, it's a long way down to the cab when you're 95 feet up underground. The ore in that "A" zone was high-grade lead-zinc. Man, it was high grade! They used to dilute it with the low grade so that the mill could handle it, otherwise it would go out in the tailings.

We miners used to think that loggers were crazy, but after the Emerald closed I went into logging. I've been disabled for thirteen years now—had a bad brain aneurysm when I was 48 and get a good pension because of it, and then later I had a bad heart attack and had three stents *[small expandable tubes]* put in my chest. I've been around a little bit, but I'm still good. I'm still young. I've been home to Nova Scotia three times this year to see my mom, who's still much alive at 85.

Canex is precious to me and I visit it every year. I know Dave Little, the guy that lives up there. There are not too many places there that you can go anymore. I don't go inside the mine, but this summer when I had my motorhome parked on the flat outside, I took my flashlight and walked in the mine a little ways.

There's an old guy who owns the bottom tailings pond four kilometres off the highway. He's the greatest old guy, lives totally independent. He has his own water plant, his own generator system. He has his own TV and everything. He grows his own food. You should see the way he keeps everything, great old guy. I go visit him now and then and you would never know a townsite had once been there. There's nothing but trees there now. Up by the Invincible, where we dumped all that waste rock from underground, there are trees growing all over as well—big trees, larch and fir that have reclaimed the land.

DALE HAZEL

I met Dale at Brien Thomas' house, where I was staying while in Salmo. He is not related to Leigh Hazel, although they both hail from Nova Scotia. Dale is now mostly employed in the logging business. He worked as an equipment operator at Canex from 1965 to 1973.

Dale Hazel from Salmo, 2007.

I wasn't a miner at Canex, I was a diamond driller, and I went there in 1965 and worked almost until it closed in 1973. My employer was Connors Diamond Drilling Co., who used to do all the diamond drilling at the mine, but the job didn't last the whole year. It would usually be for nine or ten months and occasionally eleven, so then I started filling in for Canex, running their equipment. Mostly that was driving big trucks underground. Then whenever Connors renewed their contract, I would be back drilling for them again. I did quit the mine a few times. I went to Campbell River once, and I also went on the Hugh Keenleyside dam at Castlegar in the summer of 1967. I saw an ad in the paper for iron workers, so I went over and was there from July to October. I had been an iron worker before I left Nova Scotia. I can't remember why I left the mine in Nova Scotia, but I was young and single then. At the dam I

George Murray, shift boss, and Murray Crayston, surveyor. Mid-1960s. Courtesy of John Bishop.

made three times as much as I did at the mine. I was in a big camp there with free room and board, and I worked six and seven days a week with double time for all overtime.

I came from Walton (a lot of the men at Canex came from there), a little village close to Windsor in the Annapolis valley. A whole pile of us ended up in Ymir. We were all from the same town. Some went back and some stayed. In 1965 I came to Ymir with a friend. He had come out the previous year and was home for a visit, so I came out with him when he went back. I was just looking for a job, and one of the driller's helpers had just quit, so I got in touch with him and he said, "Oh yes. I need a helper." So away I went to the mine.

There was a barite mine where I grew up, but you had to go down a shaft to get into it. I have a brother-in-law who also worked in the barite mine near Spillimacheen in BC. They still take a little bit out of there. The barite back home is a whitish colour and is shipped all over the world. They had a ball mill where they crushed the barite ore to a fine powder and bagged it. When the barite ship came in, we kids would all quit school and load the bags of barite onto the ship. It seems to me that those bags weighed 114 pounds each. That was a helluva load for a 14-year-old kid. The bags would come down a conveyor belt, down a chute onto a big long table, where we would grab them in our arms and then pile them. That gave us our spending money.

In the mine, the deeper they got the more water problems they had. Below the barite there was also a lead-zinc ore body which was never developed. When the pumps could no longer handle the water coming in from the sea, they shut the mine down. Another problem was that one of the managers had been taking too much ore and not leaving enough pillars. This led to a lot of caving. No one ever reopened the mine. Before they sank the shaft, they had mined it by open pit methods. Now there's a huge hole full of water left there.

Our drilling *[at the Emerald]* was mostly short holes, 20 to 40 feet each. They were short enough that we could pull the rods by hand. Most of my time was up in the Jersey with a little down in the Dodger. I did do some drilling in the Emerald Tungsten later on and a little down in the Invincible as well. Near the end, Connors were pretty much out and Sheppard's had taken over. I don' recall working for Sheppard's, but some of the other guys did. By then I was working mostly for Canex anyway.

The geologists would mark the holes and give us the angles and back sights for each one. It was hard work, setting up and tearing down between moves. With the holes being so short, we moved often. We had an old truck there with all the gear on it for moving. It didn't take us long. We set up a bar and clamped the drill to it, hooked up the water and air, lined up the drill, and away we'd go. We would often set up, drill the hole, and tear down in the same day. The ground was mostly pretty good drilling, usually in limestone, but we had some hard stuff too, a lot of granite, some dyke, and some skarn as well. We would burn the odd bit now and then, and get shit. We usually had a drill foreman, for there was quite a crew of us—six or eight. We worked on a three-shift schedule when we got busy. It would depend on the amount of drilling we had to do. We were usually on a contract system too. After a minimum footage we got a bonus for every foot drilled over that. It never amounted to much, but was enough to give us a little incentive. When I left the Emerald in 1972 I went over to the HB mine for Bob Logan Diamond Drilling, but I did not like it much there. I now work in the logging industry, which a lot of the Walton boys first did when they came to Ymir.

I met my wife Susan in Ymir. She was the youngest of the Ekstrom girls. We have two daughters, who have both grown up and moved to Kelowna. They have given us two grandchildren. We don't get over to see them often enough, but we do get over.

I go back to Nova Scotia quite often. My mom's still living there and I have other family there as well. I usually take a break from the logging in the spring and that's when I go back. I just fly down for a week or so. My wife goes with me too occasionally.

Reg Hallam, geologist, with his wife Chris in 1967. Courtesy of John Bishop.

Louis Lang, blacksmith. Courtesy of Sultan Minerals Inc.

B. McNeil and B. Adolphson surveying. Courtesy of Sultan Minerals Inc.

Curling team: Ian Dingwall, Bob Blackwood, John McConnell, and Lee Rowe (Ray's son), mid-1960s. Courtesy of Ross McConnell.

Lawrence Ranson, mechanic. Courtesy of Sultan Minerals Inc.

DAVE CARL LUNDGREN

I have never known a professional weightlifter, but Dave Lundgren has the build I would expect in one. Brien Thomas, who is no lightweight himself, says that Dave was the strongest man he has ever known. He displayed one example of his prodigious strength when his neighbour, Rae Thomas's station wagon slid off a snowy road and planted one rear wheel in the ditch. When Dave came along, he told Rae to get back in his vehicle. He then lifted the back end of the wagon out of the ditch back on to the road and sent Rae on his way. Dave was at Canex from 1966 to 1973.

I was born in the Nelson hospital in 1940 and grew up in Ymir, where my wife and I still live. Before working underground at the Emerald, I had previously worked as a labourer and trackman at Remac for a couple of years and at the Gold Belt for one year. At the Emerald I operated a D7 dozer and a loader for short periods as well as driving truck occasionally. That changed when "Bonus John" Wasilenko asked for me as his cross shift in a drift. The shift boss warned John that I had never worked as a miner, but that did not deter him.

John knew that the jackleg drill would be like a toy for me, so he was not concerned about my ability to drill with it. What he did was draw up sketches for me and give me written instructions each day on how to drill and load the holes for the round. He even collared the cut holes for me at first. It was not very long before I was pulling my weight and that John's faith in my ability was justified.

Much of my time at Canex was spent driving raises, including one 760 feet long driven at 57°. All experienced miners know that before starting up a raise where someone is drilling you alert them by first turning off the air briefly to stop the drill, then yelling, "Nothing down." This warns the driller that you are on your way up, and he will be careful not to let anything fall off the staging and strike you while you are coming up the ladders. On one occasion when my regular partner got sick, I had a new helper for a day. The helper had gone down for the explosives while I was still drilling. Some time later I felt a tap on my shoulder, I was so startled that I almost knocked him off

David and Diane Lundgren of Ymir, 2007. Dave is a good man to have for a friend.

Dozing underground. Courtesy of Sultan Minerals Inc.

the staging. The man had not let me know that he was on his way back up. He was also fortunate that my drill had not knocked any sizeable rocks down the raise while he was climbing up.

After the Emerald closed, I worked at the HB Mine during 1974/5. I also worked for Bob Golac once or twice around Ymir. I was at the Westmin mine at Buttle Lake near Campbell River for a while as well. Later I went to Mica Creek where I worked in an 8′-by-8′ drift until I had an accident that disabled me.

It happened when I threw the connector end of a heavy electrical cable to the fellow on the jumbo. The guy caught it, secured it, and headed down the track without looking back. Unfortunately for me, the cable had looped around my ankle. The jumbo dragged me down the track at a fast pace, bouncing me off the ties. Just before stopping, it hurled me into the air, and I landed on my head and cracked some vertebrae in my neck—an injury from which I have never recovered.

LEIGH HAZEL

Leigh was one of the many people who came from Walton, Nova Scotia, to Ymir back then. The Ymir women must have liked them, for it seems a large number of them married these guys. Leigh is one miner that I had not met or heard about until shortly before interviewing him.

I was born in Walton, Nova Scotia, in 1939. I came west and went to Western Mines, where my brother got a job. Then I returned to Ymir, where Jack Robinson hired me to work for Canex.

I was running the front-end loader in the "A" Zone one day loading two R13 Eucs that were hauling from me. We had gotten quite a bit of ore out that day, so I said to the guys, "We may as well quit a little early, go out and fuel up; we've put in a good day's work."

In the meantime, I was loading Lorne Enderburn's truck with him sitting in it, when I heard this big crash. I thought I had broken an axle or something, so I shut the machine down and asked Lorne if he heard that. He said, "Yeah, I think it's your loader." So I started the loader up and worked it back and forth and nothing happened. I said, "I might as well finish loading you, it seems

Martha and Leigh Hazel in their Ymir home, 2007.

okay." He wanted to try the loader so I let him—that's how we learned to run the equipment. Anyway, we left and went to fuel up, and by then it was quitting time.

When I came back the next morning, they had ribboned off my workplace. Over 150,000 tons [of rock] had come down exactly where we had been. That crash that I thought was the loader was a big slab falling not far from us. Where we worked, the back was ninety feet high—a big open stope held up by pillars. It had caved at the back and was working its way toward where we were. There were blocks in there the size of the Ymir Hotel right where we had been working. They had to drill and blast them to get them hauled out. Any one of them would have flattened the loader and the truck flatter than a pancake. I told the shifter, "I think I'll take the rest of the day off," and I then went home.

Another time in the A zone, I was driving truck, and I don't know if I ever told anybody about this, because at the time I thought there was no point.

I had got loaded and was at the back of the A zone. We always loaded the trucks pretty heavy; we overloaded them, you know—production and all that. I started down from the top of the ramp and was going to change gears, but I missed a gear. The adit was like a snake's back. It was all downhill through the tunnel with no place to pull off. It was just like a rifle barrel. When I missed that gear, away the truck went. The brakes wouldn't stop it. I had the emergency brake and the foot brake on, and the throttle right to the floor so that I had lots of air for the brakes. I had my hand on the air horn as I barrelled through the tunnel and came to the Wye [a fork in the tunnel]. Well, at the Wye there was usually a truck coming up and I would go down the other side into the main drift. Well, in the main drift the pipe fitters were always working, and my greatest worry was hitting their truck or the shifter, or someone else coming or going. I could have jumped out and let the truck go, but I would not do it, I stayed with it. When I got to the bottom of the main drift, straight ahead would be a big open stope which went down about eighty feet. So it was either hit that or make a left turn

Euclid truck. Courtesy of Sultan Minerals Inc.

988 Caterpillar five-cubic-yard loader. Courtesy of Sultan Minerals Inc.

towards the shifter's shack. The left turn was the only way it could go. I was carrying twenty-five tons. So I stuck with the truck and made that turn to the left past the shifter's shack. It heeled over so bad that the ore was coming off my load on one side. I thought we might roll, but it went up through what they called the "Esses" towards the mechanic's shop and finally came to a stop.

The brakes were so hot by then that they were almost burning. I sat there for a little while, I'm not sure how long, then drove out and dumped my load, but I never told anyone about how close I came. If I had hit another truck or something, we'd have been dead—at least they would have been. I might have been okay in the Euc, I'm not sure, but if I had hit the wall, I definitely would have been dead. Other than those two incidents and bits of loose falling down near me like we all experience, nothing life-threatening like that ever happened to me. Twice in six years is enough anyway.

A Koehring Dumptor hauling unit. Courtesy of Phil Steele.

Once when I was driving truck, I knocked the air line out. That was something else. I don't know how much pressure was in that line, but it had to be over one hundred psi. It let go right beside the truck window, but I had the window up and you should have seen the dirt and rocks fly, for it was an eight-inch-diameter line. As I went by it I looked in my mirror to see if the truck box had caught it, but it was okay. By the time I got down the drift a ways, one of the miners had turned off the valve. That could have cut a guy's head off.

Another time I had to go back up for something, and coming down I met two guys and a Mountie who were coming in, but I thought I was seeing things so I didn't slow down. I hit a big water puddle and splashed them.

The reason why I didn't believe it was a Mountie was that the day before we had been whooping it up in the bar pretty good and I had an awful hangover. When I came to work, coming out of the drift I thought I saw a cat. On the next trip out I saw a grouse walk across in front of me with all her chicks, so I slowed the truck down, but there was nothing there, it was just an illusion. So on this trip when I saw the Mountie, I thought it was another illusion and I never slowed down when I met them. They tried to get out of the way, but I sure splashed those guys.

JEWEL OF THE KOOTENAYS / 245

John Farkas was a mechanic, and I had just talked with him in the shop and I left. I was driving a W10 tractor trailer outfit down to the Dodger. I had hauled only about two loads when another truck driver came along and said, "Did you hear that John got killed?" I said, "John who?" He said, "Farkas." I said, "No way. I was just talking to him a few minutes ago." He said, "Well, he's deader than a doornail." I thought he must be joking with me.

John was working on the Scooptram and there was a new mechanic there—a greenhorn. John was underneath the machine and he asked the guy to start it up. The young fellow said, "No! You get out first." Now John was a big barrel-chested man and it was hard for him to get in and out so he said, "Start the goddamn machine. I can't slide in and out of here." So the guy started the machine, but it started in gear, which it never does, and it jumped backwards. John pushed to get out and he just got out far enough that his head was under the wheel, and the wheel went over his head. That Scooptram weighs about twelve tons or so, and it had steel chains on the tires. It never did start in gear, it never would—it was automatic, but somehow it started that day. Perhaps that was what he was working on, I don't know. I met a couple of guys and I said I just saw two guys and a Mountie going in there—I'm having illusions. But they had been real all right. I liked him. We used to drink beer together. He was a real good guy and a good mechanic; he was going to get married that weekend. That would have been his second marriage.

We had this mine captain, Ray Rowe, who was pretty tough, but his bark was actually much worse than his bite. A friend of mine said to him, "Ray, you are just like an old bull my grand-daddy had; one horn, one nut, couldn't fight, couldn't fuck, all he did was shit and roar." Ray didn't know what to say. He stood there momentarily with his mouth hanging open, then walked away. I got a kick out of that because everybody was a little cautious around Ray. But he never came back with an answer to that one.

Another time Ray took out a can of snuff and offered some to Alfred Gould who was a mechanic there. He had been a miner who got hurt real bad and they put him on as a mechanic. Anyway, Ray took a chew out of the can and passed it to Alfred. Alfred looked at Ray serious-like for a minute, then he said, "You know, Ray, I haven't found a hole dirty enough to put it in yet."

Ray Rowe and Roberge on a 988 Cat loader. Courtesy of Sultan Minerals Inc.

Gould got hurt pretty bad while working in a stope when a piece of loose let go. He said he heard it and knew it was coming, but didn't know where. It caught him on his hand, smashed his hard hat, and caught him in the back. When we got him outside his rubber glove was cut in two and his fingers were still in the glove. He was in the hospital for quite a while and they cut all his fingers off short, so they put him in mechanic work repairing jacklegs and such. I worked with him for years.

The mine was a great place to work. They looked after their people like an old mother hen.

PHIL STEELE

I knew of Phil's father from my years at Remac, but did not actually know any of the family personally. I met Phil when I interviewed him at his home in Salmo. He contributed a number of equipment photographs for this book. Phil worked at the Emerald from 1970 to 1973.

I was born in November 1939 at Princeton and lived at Hedley until I was eleven. My father was the assayer at the Nickel Plate gold mine. At that mine we used to say, "The coloured crew changes clothes on one side of the Dry and the regulars on the other. In the coloured crew are the Grays, the Browns, the Greens and the Whites."

Phil Steele, 2007.

Our family moved to Remac and lived above the assay office until the mine closed in 1953, when we moved to Newcastle, New Brunswick, where my father was an assayer at the Heat Steel Mine. When Remac reopened in 1955, we moved back, and my father became the mill superintendent there. My first job was pitching hay bales in Cawston near Keremeos. At Remac at age sixteen, being too young to work underground, they gave me a job driving the utility truck for Bob McDonald. Eventually I moved into the mill and worked on the Jaw and the New Holland crushers.

Later I left Remac and went to work at Canex driving equipment underground from 1968 to 1970.

BONNY KLOVANCE

Bonny was a school teacher at Salmo for many years. She occasionally filled in as a substitute at Canex and Remac. She provided the only photos of the Harold Lakes School at the Emerald I have been able to find.

Bonnie Klovance, Salmo teacher, taught as substitute at both Canex and Remac. 2007.

I was born in Duncan, BC, and completed high school there. I then took my teacher training in the Victoria Normal school. I already had my senior matriculation from Cowichan High School, which is equivalent to first year university. When they later made a degree a prerequisite, those of us from earlier times who had senior matriculation were covered by a "grandfather clause," so I did not have to go back to school, but I did so anyway in 1980 and got my degree and full qualification.

In 1952, after completing my teacher training, I went to Salmo. It was my first school. I chose Salmo over Slocan City because people in Nelson said, "Pick Salmo, the road to it is paved."

During my high-school years we had a majority of girls in the classes. While in teacher training, we had twice as many women as men, so they would bring cadets in from the Royal Rhodes military academy for our dances. That ensured there would be enough guys to dance with. When I came to Salmo, Mrs. Waterstreet was the Post Mistress. Her daughter Pam and my girlfriend were both in nurses' training at the Royal Jubilee hospital in Victoria. One day after seeing the mine buses discharging the miners at the end of their shift, I wrote to my friend in Victoria, "Salmo is really a great town. They bring the guys in here by the bus load!" My friend told Pam and she wrote to her mother what I had written. When I walked into the post office, Mrs. Waterstreet said to me, "I hear you like this town because they bring the guys in by the bus load." To think I had been trying so hard to create the "right" impression as a school teacher when I came here! I was very careful of what I wrote to my girlfriend after that.

Jim and June Grant were some of my first friends there. They were great people. She took in the youngest of the Gray boys when their parents were unable to cope because they were ill—the father later died soon after of silicosis. June Grant's own son later became a superintendent of schools in BC but now lives in Toronto. I also knew the Gray

Harold Lakes School in mid-'50s(?)—it was originally a one-room school, but was later expanded to two rooms. Courtesy of Bonny Klovance.

Harold Lakes School from a different perspective. Courtesy of Bonny Klovance.

boys and the Brown boys, the "coloured boys" we used to call them. The Dorey boys were in my class from September '52 to June '56—their father ran a garage in Salmo for many years. Then I went to Riondel and taught there from September 1954 to June 1956.

The teacher who took my job when I left Salmo was a real sergeant major in the class room. Those children had to toe the line, pay attention, and get their work done. She needed that teaching job for she had a very tough time making ends meet until her husband joined AA and went on the wagon. No one could blame her for keeping him on a very short leash until he quit drinking and turned his life around.

I had never seen the miners come out from underground until I was at Riondel, where the main road went right past the mine dry. One day I was down there when the miners came off shift. I said, "Holy Crow! I didn't know we had so many Negroes here?" I didn't realize how black the miners got at work. I'm a teacher, and I think it's so sad that the children at other mines never see how dirty their fathers get at work. If they did, they might appreciate them a lot more. Because their fathers are clean when they get home, the children have no idea of what's involved in their work unless they had bring their mine clothes home to be washed. In later years that did not happen much because every mine dry usually had its own washing machine. The wives must have detested washing their husbands' dirty diggers, for they left the washers incredibly filthy.

My fiancé Joe was working at Fruitvale when we got married, so we first lived there. We also lived for a time in Ymir, where I met Doris Thomas. She had the only phone in the village except for the one in the hotel. People would often phone and leave messages with Doris, who would then dispatch one of her many children with it to the recipient.

We lived in Balfour for a winter, but had to move again in the spring. I told my friends that we hit all of the big cities. By then my eldest child was in school, so I needed to settle down somewhere. Joe was back working in Salmo, where he had been born and where his family still lived, so that's where we moved to.

Phil Steele's wife was the secretary of the school then. She had an aneurysm on the 28th of August and passed away three days later. What a hassle. She had known how to do everything at the school. It was chaotic—the principal, who had been there for only a year, hadn't yet learned how the school ran.

The curling rink and the skating rink at Salmo were in the same building. One time the snow became so heavy on the roof that a beam cracked—it sounded like a shot going off! The skating rink was of course condemned, but for some obscure reason the curling rink was not, even though it was under that same dangerous roof. Was this because the curlers ran the town?

I used to write half a page on Salmo for the Nelson *Daily News* each week. I would sometimes tell them, "Lots of things happen in Salmo, but some I can't print."

A part of Joel Ackert's collection of mining literature. 2007.

A portion of one wall in Ackert's home museum; three walls are covered with mining photos in this room. 2007.

There were a number of people who, for one reason or another, were not interviewed or their interviews were not used. However, some of their contributions are important and are shown here.

Some of Joel Ackert's photos are shown elsewhere. Joel's father was a collector of mining memorabilia and Joel has continued in his footsteps. Here are two photos I took inside Joel's house.

John Voikin may not be much of a talker, but he like so many others, moved back and forth between Canex and the HB Mine. Despite a handicap, he still gets out on the golf course, albeit with the use of a cart. He has a son who is now a doctor and practices in the West Kootenays.

Andy Wingerak never throws anything away. I am thankful to him for being such a "packrat." He has saved old pay-stubs, utility bills, a union-negotiated pay scale and the like, which are reproduced here. I have known Andy since he and his brother Bill came to work at Remac in 1957. Andy was my cross shift for a short while in a stope, and Bill cross-shifted me in a long drift there. Bill passed away many years ago, but I visit Andy and his wife Isobel at least once a year at Kelowna. Andy worked at Canex from 1968 to 1971.

John Voikin in the new home he built in Salmo after winning a lottery. 2007 photo.

2007 photo of Isobel and Andy Wingerak at their (A&W) watering hole in Kelowna.

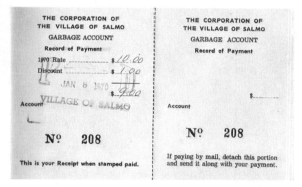

Salmo village utility bill. Courtesy of Andy Wingerak.

Pay stub. Courtesy of Andy Wingerak.

PART IV
The Dark Side of Mining

The interviews as a whole in Part III may have tended to engender an overly rosy picture. There is little doubt that life at the mine was generally good by the standards of the time, but there was a downside to it as well. The mine did on occasion exact a toll—sometimes a deadly one. Underground work was dangerous and was not far behind the logging industry in either its accident or severity rates.

There was steady, if slow, improvement in safety over the years. When the Emerald mine first opened early in the 20th century, there were almost six fatalities each year in the industry for every 1,000 people employed in it. By the early 1940s, that rate was down to two fatalities, and by the 1970s, it was even lower. Another way of assessing the danger is by tonnage mined per fatality. Figure IV.1 shows the average tons mined in the province per fatality from the early 1900s to the mid-1960s.

There were a number of reasons for the decrease in mine fatalities. In the earlier part of the century, explosives were often involved in the annual loss of life. For instance, the 1908 Minister of Mines report states that 11 of 21 fatalities were due to explosives. Only one of these was listed as caused by defective powder, but drilling into unexploded powder took three lives and blasting in general took another six.

In the 1913 report, it is mentioned that some mines are now using non-freezing powder. Apparently explosives that had first to be thawed out presented special safety problems. Over the next few decades, blasting and powder-related deaths as a percentage of all fatalities slowly diminished.

Somewhere over the years regulations concerning the amount of toxic gas output per cartridge of dynamite became an issue. Explosives

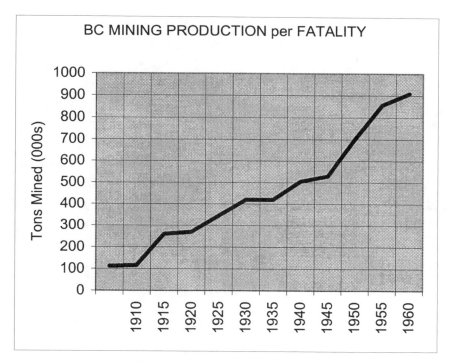

Figure IV.1.

are ranked by the amount of toxic gases released by a 1¼″ x 8″ inch cartridge of dynamite. Class 1 produces a maximum of 0.16 cu. ft. of such gases per cartridge; Class 2, up to twice that amount, and Class 3, up to four times as much. By mid century, only those explosives meeting "Class 1 fume category" requirements could be used underground.

The Emerald was unique in a way, for during its first 20 years of operation it did not have a single fatality or an accident worthy of mention in the Minister of Mines annual reports. A good part of the reason was undoubtedly that the crews tended to be small, although there were up to 40 men on the payroll at times. After Canex took over the property and greatly expanded its workforce, this was no longer true, and the company suffered 10 fatalities over a period of 23 years. Its tonnage mined per fatality was just over one million tons—slightly better that the industry average for those years. The following pages contain reprints of the BC Minister of Mines reports of the accidents causing the death of those ten men.

Mining can be an exciting business. Searching for ore may be interesting, and finding it thrilling, but there are always costs to consider. First of all, much of the work prerequisite to finding ore is tedious, painstaking work. Secondly, underground mining can require brutally hard labour. Thirdly, underground mining can be especially dangerous, for not only is the miner subject to rock falls or falling down raises, or drilling into misfired holes containing dynamite, but as the

Keith Kornum of Salmo. He mined in Yellowknife and Iceland, but only briefly at Canex. 2007.

following accounts show, miners are exposed to many other dangers found only underground. Add to this such occupational hazards as silicosis, asthma, carbon monoxide poisoning, industrial deafness, and a few others, and one may be forgiven for considering mining to be one of the most dangerous occupations. Why then do people do it? It often pays well—especially for someone who is well organized and not afraid of hard work or tough working conditions. This section documents some of the costs in human capital experienced at the Emerald while the company was wresting mineral riches from below the surface.

CANEX FATALITIES

The death of a worker by accident is a traumatic event for all concerned, but probably it is worst for the mine manager, who must shoulder the ultimate responsibility for the event. Because miners usually worked on some kind of a bonus system, they sometimes developed the habit of working hard and paying more attention to production than to safety. This sometimes led to unsafe practices, as evident, for instance, from Keith Kornum's account about the man he worked with while scaling the back from a Trump Giraffe.

"Scaling in the 'A' zone, I didn't care for at all. We had a brand new cherry picker, *[Giraffe]* I think it would go up 90 feet and you'd still have to stretch and reach and reach to get the big loose down. My partner was a guy who feared nothing. He'd get up and stand on the top rail of the cage and be barring, like that's 90 feet off the ground! And then he'd get the bar on a rail and use it for a pry. I got him to take me down and I said, 'Thanks a lot Bud. I'll see you.' I said, 'I'll stand here and watch and make sure you don't get hurt.'"

There were ten men who paid the ultimate price at Canex. The following accounts are taken from the Minister of Mines Annual Reports for the years in which they occurred.

1947

On November 17, 1947, August Wegener, miner, was instantly killed in the 3970 underhand stope at the Emerald mine near Salmo. Wegener

Table IV.1: Canex Fatality Summary

Year	Victim	Occupation	Age	Cause
1947	August Wegener	Miner		Buried in ore pass
1949	Nelson Earle Roynon	Ball mill Op.		Fall in the mill
1951	George Dickson	Miner	26	Buried in stope
1952	Leonard Oliver Gostling	Switchman	56	Crushed between chute and car
1952	William Douglas Bauman	Miner	23	Fell down 3865 ore-pass raise
1956	Alex. S. Chernoff	Electrician	44	Electrocuted
1958	John Daniel MacDonald	Slusherman	52	Hit by rolling boulder while slushing
1964	Wesley Schneider	Miner	42	Choked by rag around neck while drilling
1966	Wayne Ake Johannessen	High Scaler	27	Scaling off Giraffe when it collapsed
1967	John Farkas	Mechanic	45	Crushed under scooptram

and his partner were drilling large slabs of broken ore over a chute. The blasted material had not settled solidly into the chute, and the water and vibration of the drill caused it to subside suddenly into a recess in the wall. Wegener was taken down with the muck and was covered by material which slid in after the subsidence.

1949

On March 20, 1949, Nelson Earle Roynon, rod-mill and ball-mill operator, died as a result of injuries received when he fell from a platform to a concrete floor in the Emerald mill of the Canadian Exploration Company. The platform had satisfactory railings, but the evidence would indicate that the deceased sat where he could watch operations and dozed off and fell about 15 feet.

1951

On September 28, 1951, George Dickson, Canadian, aged 26 years, married, employed as a miner by Canadian Exploration Limited in the Jersey mine, was asphyxiated in No. 4044A4 stope when he was buried under muck which suddenly settled after being hung up. Dickson and his partner had been detailed to drill holes for the purpose of slashing the top of the raise preparatory to installing a grizzly at the proper level. As the muck was too high to allow drilling in the correct place, the shiftboss arranged to draw a train load (ten 2-ton cars) from the chute below, after which he was to return to the top of the raise to detail the work. Six cars had been pulled when a spill occurred, and a short time later the ore was heard to settle in the ore-pass. Dickson's partner called down the manway that Dickson was buried. When his body was recovered about thirty minutes later, there was no sign of life, the deceased having died from asphyxia and shock. Evidence submitted brought out the fact that Dickson and his partner noticed the muck settle in the raise and then stop. They suspected a spill had occurred but did not know how many cars had been pulled nor did they expect a hang-up. They knew, however, that the muck had not gone down the equivalent of ten cars. **Being on contract, they were anxious to start work** [my emphasis]. They started to collar holes about 4 feet apart when the muck subsided, burying Dickson.

1952

Leonard Oliver Gostling, Canadian, aged 56, widower, and employed as a switchman by Canadian Exploration Limited at the Emerald mine, near Salmo, was instantly killed when he was crushed between a car and chute at the 3968 stope on 3900 level dump, January 11th, 1952, at 1.05 p.m.

The 3968 stope is drawn by a slusher, set up on the first lift (or 6 feet) above as 3900 tramming level. The ore is scraped through a chute directly into the cars.

One car of a five-car train was loaded and, while loading the second, the body of the car, a gondola side dump, became knocked off its rocker. The car was righted by W. Frame, motorman, and M. Maxsemenko, car-spotter, after the three empty cars had been disconnected from the train and moved a car length north up the track.

While this was being done, Gostling arrived on the scene, possibly to ascertain when the motor would arrive at his working place, which was nearby. He did not assist righting the car, except to pass over a pinch-bar, but remained standing beside the south side of the chute between the car and the wall timber and facing the motor.

After the car was righted Frame returned to the motor to move it to connect up with the three disconnected cars. He was facing Gostling, who was standing in the same position beside the track not more than 3½ feet away. Frame said "Let's go," but then he inadvertently moved the motor the wrong way to the south about 2 feet. He immediately reversed the motor, and when he looked up from the controls as the train moved north, he saw Gostling pinned between the car and the chute lip. He stopped immediately, but by then the train had moved 4 feet. Maxsemenko had just placed the pinch bar against the side of the drift, and, as he turned, saw Gostling's right shoulder hit the south lip of the chute and his body make a half turn as it rolled between the car and the chute lip.

Gostling was pinned by the chest, face up, and the car had to be dumped to release him. There was no sign of life, but help and first aid was on the scene almost immediately. The doctor arrived about 2 p.m. and pronounced the man dead.

The inquest was held in Salmo on January 16, and the following verdict returned:

"L. O. Gostling met his death at approximately 1.05 p.m., January 11th, 1952, on the 3950 level at the slusher chute in 3968 stope of the Tungsten Mine, Canadian Exploration Limited, situated approximately 10 miles from Salmo, BC. Death occurred in an accidental manner by Gostling being rolled, crushed and pinned between the lip of a muck car and the lip of the chute as the train was being moved north. We find no evidence of negligence on the part of the company or the workmen involved in the accident."

It is difficult to find an explanation for this accident. He was an experienced man and was well aware of what the train crew were doing. He might have been surprised when the train first moved south instead of north and could have leaned over the cars to see why and so was caught when the motor was reversed.

William Douglas Bauman, aged 23, Canadian, married, employed as a miner at the Emerald Tungsten mine of Canadian Exploration Limited near Salmo, died as a result of injuries received when he fell down 3865 ore-pass raise at about 12:40 p.m. on December 3rd, 1952. The 3868 raise connects the 3868 sublevel with the 3950 main haulage drift. It is driven on a slope of about 55 degrees, is about 5 by 5 feet in cross section, and has two knuckle-backs in its length. It was to be used as an ore-pass

The deceased, together with another miner, Ralph McKeown, and a timberman, Frank Dryzmala, were working, at the top of the 3868 raise on the 3950 level preparing the location for a grizzly. During the first part of the shift they had been drilling hitches from a platform about 5 feet below the 3950 level. This platform consisted 2-inch planks resting on two 4 by 6-inch timbers 5 feet apart hitched into the sides of the raise. At lunch-time the planks were removed and the holes blasted. When the men returned, they cleaned around the opening, Dryzmala working on one side and Bauman and McKeown on the other, Bauman then decided to descend the ore-pass to see if the two sprags were still secure enough to be used for the planking which had to be replaced as a safety measure. He descended the raise, facing the footwall while holding on to a rope held by McKeown. Apparently, as he placed his weight on one of the sprags, the support gave way suddenly and he lost his grip on the rope and fell down the raise a distance of about 70 feet.

Help was obtained immediately, and Bauman was removed to the first-aid room in a basket stretcher. He was still breathing faintly at 1:10 p.m., but was pronounced dead by the doctor when he arrived at 2 p.m. An investigation revealed both the sprags were sound; the rope was in good condition although slippery, and a safety belt which the shiftboss had provided, was nearby. The use of this belt very possibly would have prevented the fatality.

An inquest was held in Salmo on December 10, and the jury returned the following verdict:

"William Douglas Bauman met his death on December 3rd, 1952, at approximately 1:45 p.m. in the tungsten property of the Canadian Exploration Co., with no blame attached to anyone. Bauman fell down 3868 ore pass from the 3950 level while working in the grizzly opening. The jury recommends that better safety precautions be taken when work of this kind is being done."

A safety belt similar to the one provided was tested and found to have adequate strength. All men engaged in the work around the grizzly were experienced.

1956

Alex. S. Chernoff, aged 44, Canadian, single, and employed as a first-class electrician by Canadian Exploration Limited, Salmo, was apparently instantly killed when he was electrocuted at No. 410 underground substation in the Jersey mine on July 26, 1956, at about 11.30 a.m.

Chernoff, in company with Earl Gilbert McLean, electrical foreman, arrived at the transformer-station about 11.25 a.m. A combination magnetic starter, formerly operating a fan motor, was to be removed for use elsewhere. Energy is supplied to this starter from a 440-volt 3-phase distribution panel containing five breaker switches. McLean lifted the lid and observed they were all in the "off" position except the top left-hand one, which controlled the lights in the transformer-station and near-by workings. McLean turned this switch off and then on again as a check of the "on" and "off" positions. (There is a certain amount of confusion with this type of distribution panel as the upper switches are "off" when up and the lower one "off" when down.) McLean lowered the panel cover and told Chernoff, who was beside him, to proceed with the removal of the starter. He then left the transformer-station to investigate a raise near by which would be their route of travel. Returning in about two minutes he found Chernoff lying on his back, his left hand grasping the end of three wires which he had apparently just removed from the starter box. McLean looked into the distribution. panel and found the lower right-hand switch controlling the electricity to the starter in the "on" position. He turned it off and removed the wires from Chernoff. Help was obtained and artificial respiration applied until the arrival of Dr. Carpenter from Salmo about 1:10 p.m., who pronounced the man dead.

The deceased was an experienced, qualified electrician and had worked for Canadian Exploration Limited for the past five years. There is no explanation as to why he turned on the breaker switch after McLean left, unless he checked the position and became confused. He had a tester with him but apparently did not use it.

The autopsy showed an abrasion over the left eye and deep electrical burns on both hands. The doctor, after listening to the evidence, stated he believed that death was due to respiratory failure.

An inquest was held at Salmo at 7:30 p.m. on August 1st, 1956, and the Coroner's jury returned the following verdict:

"We, the jury, find the deceased Alex. Sam Chernoff met his death by electrocution at the 410 substation located in the 4200 level of the Canadian Exploration Jersey Mine on July 26th at approximately 12 noon, 1956, electrocution due to the deceased's misjudgment of the switch controlling the current which fed the box on which he was working. This jury recommends that a more visual identification be used to determine the "off" and "on" position of the switches in these darkened areas."

The recommendation of the jury is agreed with. The words "On" and "Off" cannot be observed in this type of distribution panel after it has been in use underground for a short period. The difference in the "off" positions of the upper and lower switches is most confusing to a layman, and apparently in this case to a qualified electrician. The electrical superintendent of Canadian Exploration Limited pointed out that this type of distribution panel in use in Japan had installed in it little neon lights for each switch. He suggested that this might be incorporated in Canadian models. Another suggestion is that part of a switch be recessed behind a sliding door, which would have to be lifted to put a switch in the "on" position. The distribution panel in question had been approved by the Canadian Standards Association, and it is therefore recommended that the above suggestions be brought to the attention of that association.

1958

John Daniel MacDonald, aged 52, Canadian, married, and employed as a miner at Canadian Exploration Limited, Salmo, was killed by a rolling rock in the 3556 scram drift in the Emerald mine on July 27, 1958 at about 11:35 a.m.

The 3556 scram drift rises to the north for about 150 feet at an angle of about 25 degrees. An open stope, above and to the east, provides the ore for the scram drift. A 10-horsepower air slusher hoist is set at the south end of the drift, and it moves a 36-inch scraper down the slope to the grizzly in front of the hoist.

There were no witnesses to the accident. MacDonald was operating the slusher at lunch time and when his partner returned one-half hour later, MacDonald was found slumped over the controls with the motor running. The top of his skull was crushed. A disk shaped rock weighing about 400 pounds was lying at his side. It was assumed the rock tumbled into the scraping channel and then rolled down the slusher trench, gaining sufficient momentum to continue across the grizzly to strike MacDonald with considerable force.

An inquest was held at Salmo on July 31 and the coroner's jury returned the following verdict:

"We the jury, duly sworn and empaneled July 28, 1958, to enquire into the death of John Daniel MacDonald do find that the deceased John Daniel MacDonald of Salmo, BC, came to his death from a fractured skull caused by a flying rock received while working underground at the Emerald mine. He died between 11:35 a.m. and 12:05 p.m. July 27th 1958. We feel that no blame is attached to anyone. We do however feel that the Company management give consideration to placing of guards in front of the slusher hoists operating where scraping angle is over 15 degrees."

No previous incidents of rolling rock had been reported for this scraper drift, and thus the hazard had not been foreseen. However, it is apparent some sort of guard might have halted or diverted the rock, thus the verdict is concurred with. It has also been recommended that for similar set-ups the slusher hoist could be located in an offset or less exposed position.

1964

Wesley Schneider, aged 42, married, and employed as a miner at the Jersey mine of Canadian Exploration Limited, was choked to death while operating a drilling-machine about 5:15 p.m., September 23, 1964. Schneider was working alone on the afternoon shift in the 72G stope. This stope is flat lying, 180 feet in length, and varies from 15 to 65 feet in width. The ore is removed by slusher to a scram drift at one end of the stope. Schneider received his instructions from the underground shifters' office at about 4:10 p.m. It would appear he then proceeded to the stope, drilled one 61/2-foot hole in the back of the stope with a jackleg machine, and then started a second hole when the accident occurred. Schneider was apparently holding the jackleg with one hand while guiding the drill rod with his right hand, thus bringing his upper body in close contact with the drill rod. The bit had a tendency to run downwards while collaring, so Schneider may have been supporting the drill steel with his chest. A neck band made of waste rag, which Schneider was wearing, became attached to the rotating drill steel and caused death by strangulation. The deceased was found lying face downwards across the drill steel. The rock drill was set at about one-half throttle but was not running properly due to rotation being stopped.

At about 5 p.m. an electrician entered the stope to repair Schneider's 20 horsepower electric Blusher, which was about 110 feet from the scene of the accident.

Around 6 p.m. the electrician finished his job, flashed a floodlight in the direction of Schneider and left the stoping area. The electrician heard a machine operating but saw or heard nothing unusual.

At about 6.45 p.m. the shiftboss was making his regular rounds and came upon the scene. He summoned help from a nearby stope, and artificial respiration was started and kept up until the doctor arrived and pronounced death at about 8.30 p.m.

The inquest was held in Salmo on October 1, 1964. A verdict of accidental death was returned with no blame attached to anyone. There was a recommendation "that the wearing of sweat bands at the mine be prohibited."

1966

Wayne Ake Johannessen, aged 27, married, and employed as a miner at the Jersey mine of Canadian Exploration Limited, was fatally injured on May 27th when a scaling-platform on which he was working collapsed.

The accident occurred in a large open stope where mining is by room-and-pillar method and using trackless diesel-powered equipment. The scaling-platform, a "Trump Giraffe," was mounted on a Dart truck. The essential parts were two booms, about 20 feet long, hinged together and capable of being raised or lowered by hydraulic controls on the platform attached to the end of the upper boom.

On the day of the accident, Johannessen and a partner were using the scaling platform to bar down a wall of a stope. At about 5:05 p.m. the upper boom knuckles, fitting around the shaft between the upper and lower boom, failed, dropping the upper boom instantly and causing the men to fall onto a pile of broken rock 20 feet below. The noise was heard by men in a nearby stope and rescue got under way shortly after. An ambulance transported the injured men to the Salmo clinic, arriving about 6 p.m. Johannessen's partner was detained there to be treated for bruises and head cuts, while Johannessen was taken to the Kootenay Lake General Hospital at Nelson. He died at 9:45 p.m. Causes of death were internal injuries coupled with severe concussion.

The scaling-platform, designed for a maximum of 1,500 pounds on the platform, was at time of failure carrying a load of about 450 pounds. It was 15 years old and had been in fairly constant use. Investigation disclosed that the failure at the knuckles of the booms was where two of the original welds had parted. There was no evidence that any large pieces of rock had struck the "trump giraffe."

The inquest was held in Nelson on June 1st. The jury returned a verdict of accidental death with no blame attached to anyone, but added a rider recommending that machines of this nature should be dismantled and thoroughly inspected periodically.

1967

John Farkas, aged 45, married, and employed at the Jersey mine of Canadian Exploration Limited, was killed instantly when his head was crushed under a wheel of a Scooptram on November 29.

On the day of the fatality, Farkas was repairing the starter of the Scooptram and was assisted by an apprentice who was inexperienced with the Scooptram. Farkas was lying face up on a creeper under the back end of the machine and asked the apprentice to try the starter. He assured the apprentice that it was safe to do so. The machine started successfully, but as the throttle was stuck at about three-quarters of full throttle the engine, being in first gear of reverse, developed sufficient power to overcome the emergency brakes. As the Scooptram started to move back, Farkas tried to get out but his head was crushed under a wheel.

A throttle return spring had been broken previously and was unreported by the operators.

The inquest was held in Nelson on December 13th. The jury found no blame attached to anyone, but added a rider stating that all machine defects be reported in writing and that repairs be effected without delay.

THE ADIN TETZ INCIDENT

Not all accidents end in death though, and the following is the story of a horrific ordeal following an accident.

Thirty-two-year-old Adin Tetz was working for Canex as a miner on August 14, 1955, when he was knocked down by a large rock slabbing off the face in the Invincible adit in which he and his two partners were working. He was pinned by the rock that had crushed one foot, broken his thigh, and almost severed one arm at the elbow. Adin was bleeding, and when his co-workers, Walter Panagopka and Pat O'Connell, were unable to budge the rock pinning him, he ordered them to reach into his pocket for a jackknife and complete the amputation of his arm. He was convinced that he would bleed to death if he was not promptly freed, an opinion concurred with by Doctor Carpenter, who later treated him.

Here the stories diverge. The official account and the news media had it that his partners completed the job, but his close friends and his son Terry disagree. They told me that neither Pat nor Walter were able to bring themselves to complete the grisly chore, and that Adin finished the job himself. His niece Sharon Tetz says that it was his concern for leaving his six sons fatherless that

The Tetz family. Courtesy of Garnet Tetz.

Dina and Adin Tetz. Courtesy of Garnet Tetz.

JEWEL OF THE KOOTENAYS / 263

motivated him to carry out this gruesome and difficult task. Since Adin and his co-workers are no longer alive, we can never know for certain which version is correct. However, the important part of the story was not who did it, but what Adin Tetz achieved afterwards.

Adin was fiercely independent. Not only did he become a productive welder, but he learned to tie his own shoelaces and roll cigarettes with one hand. He spent the rest of his days in severe pain and when the morphine was cut off, he turned to alcohol to deaden it. Despite his problems with liquor, the company management thought enough of him to not only provide a second mortgage for the house he built at Williams Lake while he was working at their Gibraltar mine, but to forgive that mortgage when he died on December 11, 1973, at age 50.

Here is what two of Adin's sons remember:

Adin Tetz welding. Courtesy of Sultan Minerals Inc.

GARNET TETZ

I met a guy about ten years ago who said, "I knew your dad when he worked at the Emerald mine. He was welding and he had this wheel barrow that was full and he had no way to move it. He just reached down and grabbed one handle and looped it over his shoulder, then grabbed the other handle with his remaining hand and wheeled it away. It was the first time I met him and it was something I never forgot."

Dad was stubborn and independent. He did learn how to roll a cigarette and tie his shoelaces with one hand, but he did not want to do the dishes. He just didn't do dishes, but there was nothing else he didn't do. I never thought of him as a one-armed man, for if there was a hole to be dug or something to be welded, he just did it.

TERRY TETZ

My dad worked for Placer at the Gibraltar mine for some time after leaving the Emerald. He built a house in Williams Lake and got a second mortgage from Placer. I remember that after he died, the mine

GIBRALTAR MINES LTD. (N. P. L.)

August 30, 1974

Mr. J. M. Gibbs
Mine Manager
Gibraltar Mines Ltd. (NPL)
Box 130
McLeese Lake, B.C.
V0L 1P0

Dear Mr. Gibbs:

<u>Re: Mrs. Adin Tetz</u>

It has come to my attention that Mr. A. Tetz died in Williams Lake Hospital in December 1973 following complications as a result of an operation.

For the record it should be remembered, as I recall clearly, Adin Tetz was employed at the Canex Operation at Salmo some 20 years ago. At that time he was recognized as a good hard-working responsible miner and was working in the Tungsten Shaft where he lost his arm in a fall of rock from the face of the incline under rock-burst type conditions. For some people this accident would be the end - but such did not apply to Adin Tetz. After his recovery he went to trades school, completed a welding course, and continued working for Canadian Exploration and subsequently for Gibraltar. As a welder he continued to show those commendable personal characteristics of a miner, that is, despite the loss of his arm he was a cheerful, hard-working, competent, responsible employee and a credit to himself and the Company.

As I understand the situation, prior to his death Mr. and Mrs. Tetz built a home in Williams Lake subject to a second mortgage provided by Gibraltar. It is my strong suggestion, in view of Mr. Tetz' long and responsible employment by the Company, that these circumstances be recognized by Gibraltar through forgiveness of the second mortgage.

It would be quite in order for you to send a copy of this letter to Mrs. Tetz, together with your own letter, advising her of the action taken on the mortgage. I understand Mrs. Tetz is leaving the Williams Lake area in the very near future. Please express my condolences on the passing of her

700 BURRARD BUILDING • 1030 WEST GEORGIA STREET • VANCOUVER, B.C., CANADA V6E 3A8 • (604) 682-7082

Doug Little letter forgiving mortgage. Courtesy of Garnet Tetz

forgave that mortgage without being asked to. The company really looked after its people.

Dad was a real fighter. It was many years before he told me everything about losing his arm in the mine. He was pinned under a huge chunk of rock and slowly bleeding to death; it was not his partners who cut his arm off to save him. The rock had almost severed the arm and there wasn't much left holding, but he wound up having to finish the job himself. He had told his co-workers, Wally Panagopka and Pat O'Connell, to go into his pocket and get out his jackknife and cut through what was left. One of them had begun the job, but couldn't do it. Then the other guy tried and couldn't bring himself to do it either. So Dad said, "Give me the knife," and then finished the job himself.

People may not realize it, but when my dad passed away, it was a direct result of his accident. The people at WCB knew that, but they told us that there would be no widow's pension. We went to a lawyer and to a guy in the WCB. We worked on it for years and realized that Mom should have gotten a pension. The WCB was going to write her a cheque until they realized how much it amounted to with compound interest. In the end they turned down her claim because she didn't apply within the mandatory time period. Of course, the reason she didn't apply in time was because they had told Dad that when he died, so would his pension. We actually did have a letter proving we had tried to apply, but we had lost it. We fought the case for many years, but gave up after Mom died. The people at WCB knew they were at fault, but when does someone in government ever own up to a mistake?

Gil Mosses. Courtesy of Sultan Minerals Inc.

THE GIL MOSSES RESCUE

Sometimes accidents resulted in heroic rescues and brought out the very best in the people involved. This was certainly the case with the Gil Mosses rescue, where twelve rescuers received awards for bravery from both the WCB and the Andrew Carnegie Foundation. Each award, in addition to a handsome medal, also included a sizeable amount of money. There was only one Carnegie medal, but the WCB awarded gold, silver, and bronze medals as well as parchment awards, and these were accompanied by sums of $2,000, $1,000, $750, and $500, respectively. Those were fair chunks of change back in 1969 when a miner's wages were approximately $3.00 per hour plus whatever bonus he earned. This was also the first time that the WCB issued a gold medal, and Carl Shelrud was the recipient for his role in the rescue.

This is how the Workmen's Compensation Board sums up the incident:[15]
The dramatic rescue operation began about 2:15 p.m., April 18, 1969 when truck driver Jim Wiewor found the ore-pass he had been hauling from in the underground operation empty, and decided to check number 65 to see if it had been cleared.

He positioned his unit under the unloading chute and prepared to operate the controls that would allow the tons of ore to pour into his huge Euclid dump truck to be hauled from the tunnel for processing. For some reason, unknown even to himself, Wiewor called out to see if the area was clear, an unusual thing to do as normally the truck driver is the only one in the area of the chute.

Much to his surprise, he heard a voice from behind the closed gates of the chute. Realizing the seriousness of the situation, he ran to the nearest mine phone and called for help. A rescue team of Mosses' fellow workmen was quickly assembled to begin the long, dangerous job of freeing him.

It was impossible to determine the exact location of the victim because he was behind the timbers and gate holding back the tons of rock. To open them would have dislodged the entire load of ore which would surely have crushed him.

On inspection, the shaft proved to be completely full of broken ore, which would be impossible to remove from the top, so the rescue team had to face the highly dangerous task of approaching the trapped man from the bottom of the chute without disturbing the precarious position of the rocks under which he was buried.

It was decided to use an air powered chain saw to cut through the upper restraining timbers of the chute bulkhead and gingerly work toward the trapped man.

An opening about two feet square was provided and after removing several of the large rocks and the bulkhead, Mine Manager Edward Lawrence

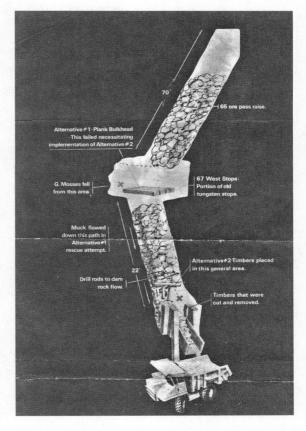

Artist's sketch of the ore-pass where Gill Mosses was trapped. Note that the truck is not drawn to the same scale as the ore-pass—from World of Placer *newspaper, December, 1969. Courtesy of Dale Burgess.*

---
15 *Courtesy of Debby Swedberg.*

and Shiftboss Carl Shelrud were able to carefully crawl into the area where Mosses was trapped to evaluate his condition and determine what would be required to prepare him for the long ordeal that was to follow.

A plan was evolved whereby it was decided to try and place timbers to prevent the ore in the upper 70-foot section from coming down, and then carefully draw the remaining 22-foot portion of the broken rock past the trapped man.

This plan worked well for a time and about 25 tons of ore had been removed when suddenly the timbers over the junction area collapsed from the weight of the rocks above.

By this time, Mosses had been trapped approximately seven hours.

The only alternative now was to attempt to place timbers against the "hung up" rocks in the lower portion of the ore-pass and carefully remove the material trapping Mosses.

The shoring was completed and removal of the rock began about 9:30 p.m. Progress went well for the first half hour and by 10:00 p.m. the trapped man was free down to his knees. However, the rescuers were now faced with a large boulder under his knees which had to be carefully broken so as not to disturb the "hung up" rocks supported by the timbers. To accomplish this, the rescue team had to take turns lying on their stomachs in the crawl space and carefully use a hammer and steel to chip the unseen rocks, all the time facing the threat of rock-fall, which would have surely meant serious injury or death.

By 11:30 p.m. several more large rocks were discovered closely packed around the trapped man's ankles. This problem seemed almost insurmountable because of the distance the rescuers had to reach under the man to work.

The continual pounding had loosened the supporting timbers and these also had to be reset.

At this point, it appeared that amputation of Mosses' legs might be the only way to free the victim, but the rescuers persisted and at about 12:50 a.m. Shiftboss Shelrud reported he was able to free one of Mosses' ankles. By 1:00 a.m. the second ankle was freed and the man was carefully lowered down the chute into a stretcher, and by 1:15 a.m. he was on his way by ambulance to the Kootenay Lake General Hospital at Nelson.

At the hospital, Mosses was found to be suffering from shock and bruises in addition to several fractured ribs, a miracle considering that after attempting to dislodge the hang up in the upper 72 feet the ore shifted, taking him down the remaining 22 feet of the ore-pass.

He is now recovering and hopes to return to work in early September.

1965-6 Industry Mining Fatalities

Mine or Place	Location	Number of Fatal Accidents 1966	Number of Fatal Accidents 1965
Boss Mountain	100 Mile House		1
Bralorne	Bralorne		1
Britannia	Britannia Beach	1	
Brynnor	Ucluelet		1
Caledonia	Retallack	1	
Cariboo Gold Quartz	Wells		1
Coast Copper	Port McNeill		1
Craigmont	Merritt	1	
Giant Mascot	Hope		1
Glacier gulch	Smithers		1
Granduc	Stewart	1	
Grouse Creek (placer)	Wells		1
H.B.	Salmo		1
Hecla	Silverton	1	
Jersey	**Salmo**	1	
Mineral King	Toby Creek	1	
Mount Washington	Courtenay	1	
South Gold	Stewart		1
Totals		8	10

Table IV. 2. From 1966 Minister of Mines Annual Reports for British Columbia.

ACCIDENTS CAUSING DEATH OR INJURY*

CLASSIFIED AS TO CAUSE		
Cause	Number of Accidents	Percentage of Total
Atmosphere	2	0.5%
Explosives	9	2.3%
Falls of ground	70	18.1%
Falls of persons	88	22.7%
Lifting and handling material	36	9.3%
Machinery and tools	87	22.5%
Transportation	18	4.7%
Miscellaneous	77	19.9%
Totals	387	100.0%

CLASSIFIED AS TO THE OCCUPATION OF THOSE INJURED		
OCCUPATION	Number of Accidents	Percentage of Total
Underground		
Chutemen	6	1.6%
Haulagemen	18	4.7%
Miners	112	28.9%
Helpers	8	2.1%
Timbermen	11	2.8%
Mechanics, electricians, etc.	10	2.6%
Miscellaneous	13	3.4%
Surface		
Mechanics, electricians, repairmen	41	10.6%
Mill and crusher workers	24	6.2%
Carpenters	5	1.3%
Miners and drillers	48	12.4%
Vehicle drivers	31	8.0%
Miscellaneous	60	15.5%
	387	100.0%

Table IV. 3. From 1965 Minister of Mines Annual Reports for British Columbia.

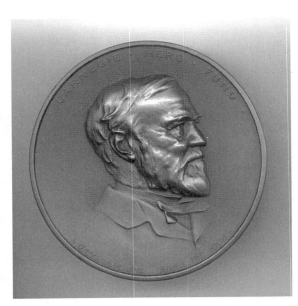

Carnegie Medal. Courtesy of Dale Burgess.

We also have an eye witness account by James (Dale) Burgess, a colleague of Gil Mosses:

I was on shift when Gil Mosses got trapped in the chute. During the rescue operation we found his leg was pinned by a big rock. The doctor wanted to cut it off to free him, but we said, no way, let us keep on working and we will get him out of there. We knew that when we got to him, half the battle was over with. We couldn't see him for hours and hours until we dug our way in to him and then they were talking about amputating his legs. We didn't care who said what, or anything. That was the way we worked and we got Gil out of there without cutting his legs off. He had fallen, hit the switchback and rolled and he was lying there head downwards buried in muck. He was hurt before the ore had buried him. The hung-up ore that he was going to blast came down after he fell and buried him. The doctor couldn't get anywhere near him until we had cut the chute out and scraped out a lot of muck by hand. When night shift came on we stayed. We were not going anywhere. I think it was 1:30 that night when we got him out.

Sonny Burgess (no relation) and Carl Shelrud were the shift bosses, and between them they made the decisions on how we were going to rescue Gil. Wayne Ritter and I were partners and took turns going in there pulling the muck down with our bare hands. We'd have a belt on with a rope attached. If the muck let go they could drag you out onto the deck as you came, otherwise you'd go into the truck and be buried too. The raise was probably six feet wide by six feet deep, so there was room for only one guy up there at a time.

I can remember cutting those chute timbers and expecting to see blood on the saw chain—we didn't know where he was, no idea. He was in the switch back, and it was lucky he wasn't any higher, for we would have had to draw more muck out to get him down. It is amazing the man lived. He fell, hit the switchback, and 1,200 tons came down on top of him. We probably pulled a hundred tons out getting to him.

Gil didn't walk for a long, long time, but he eventually came back to work. I remember when we brought him out. I drove the Dart with all the guys on it, and we sang on the way out. That's how happy we all were.

It had to be an unbelievable ordeal for Gil. He was lying there in the

Canex/Salmo Heroes Receive Awards For Bravery

Workmen's Compensation Board of B.C. "Bravery Award" recipients are shown following presentation ceremony in Trail, B.C., August 18, 1969. The twelve risked their lives to rescue Gilbert Mosses, 53, who was nearly completely buried by 300 tons of broken ore at the bottom of a 92 foot ore pass at the Jersey Mine Canex/Salmo, last April (see last edition). The WCB "Bravery Awards" consists of the $2,000 Gold Award, $1,000 Silver Award, $750 Bronze Award and $500 Parchment Award. **Recipients are (foreground, left to right) Edward Gladu, Bronze Award, Alphonce Grotkowski, Bronze Award Carl Shelrud, Gold Award; Laurent Heroux, Parchment Award; Dr. Ian Stewart, Bronze Award; (centre) Edward Lawrence, Silver Award; Dale Burgess, Bronze Award; Brian Martin, Bronze Award; (back row) Andrew Burgess, Silver Award; John Voykin, Bronze Award; Wayne Ritter, Bronze Award; and Graham Bingham, Bronze Award.**

Mr. Carl Shelrud, recipient of the "Gold Award," the highest award available, is the first person ever to receive a WCB "Gold Award."

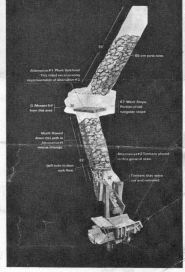

Sketch shows mine area where dramatic rescue of Gilbert Mosses took place, on April 18, 1969, at the Jersey Mine – Canex/Salmo.

PAGE 6 — THE WORLD OF PLACER — DECEMBER, 1969

Twelve Canex-Salmo Heroes. World of Placer, December 1969, Courtesy of Mel Olson.

rock for twelve hours, upside down, with the cold water running down on him, not knowing whether or not we would get him out. Old Gil and I were good friends.

Ed Lawrence, the mine manager at the time, was present and involved in the rescue. He said that after they finally freed Gil and got him on his way to the hospital, the men were so overjoyed that they all sang as they rode out to surface in the back of an ore truck. Outside, during the rescue operation, it turned out that many people had gotten together and held a prayer meeting for Gil as well.

HEALTH AND SAFETY

Mine safety in the twentieth century advanced slowly through improvements in explosives, equipment, legislation, and safety training. Provincial regulations gradually increased and required ever stricter compliance. Reducing silicosis-inducing dust was one important step. Even though dry drilling was illegal, many miners were slow to comply, preferring to leave the water off, at least while collaring a hole. Many did not take the time to wet down the muck piles properly before handling the broken rock because "time was money," and it reduced their production and hence their bonus earnings.

Hard hats and hard-toed boots have long been mandatory, and since the hat is required for carrying the miners' lights, there never was resistance, nor with boots, because a man would not be allowed underground without them. Safety glasses and goggles are a different matter. There is little doubt that they reduce eye injuries, but they tend to become spattered with mud and are a genuine nuisance to wear.

The Mines Act requires monthly safety inspections, but in the past many miners felt that their concerns were not properly dealt with if they cost money. Serving on a safety committee cuts into the contract earnings, and hence the more experienced miners did not like to do this (they earned only regular hourly wages for time spent on inspection). So relatively inexperienced miners were often assigned to this duty.

Ventilation requirements were gradually increased over the years. The allowable amount of carbon monoxide in the mine atmosphere has been cut drastically since I quit mining in 1964, probably by a factor of 10.

Improved technology has increased the use of rock bolts and wire mesh used to stabilize the backs in work areas and adits.

The use of Thermalite igniter cord has removed the tension and haste inherent in the lighting of tape fuses for a blast. The following paragraphs explain why this cord was eagerly adopted by miners.

BLASTING WITH TAPE FUSE

A development miner's work usually consists of driving a drift or a raise in order to prepare a block of ore for production mining. In my case, this was usually a 5′ x 7′ adit, which required 26 drill holes, 21 of which were loaded with explosives.

In many of my years as a development miner I usually blasted my rounds using tape fuses to set off my primers. The most important aspect of fuse length is the length of the drill hole plus an allowance for trimming plus an allowance for the time required for the miner to reach a safe place after lighting them and before the first shot fires. My fuses were typically three to four feet longer than my drill holes, for a total of 12 feet.

I loaded each hole with the primer at or near the bottom of the hole, and then filled the balance of the hole with sticks of dynamite and/or wooden spacers as needed. In order to blast these loaded holes in the desired sequence, I trimmed each fuse, taking the longest trim from the hole to be fired first, making it the shortest fuse; then the second longest trim from the hole to fire next, etc. Where the sequence between several holes did not matter, the fuses were trimmed to the same length in order to conserve the amount of trim required. Just prior to lighting the fuses, I would first split the end of each fuse open with a sharp knife to make the job of lighting it easier, and then with a "hot-wire lighter" (similar to a sparkler used on a birthday cake) I lit the fuses in the same order as I wanted them to fire.

This may sound easy and simple. It was not. The burning fuses created lots of dense smoke, making it difficult to see which fuses had been lit and which had not. Sometimes water might drip on them, and I would have to cut off a short piece of fuse to create a dry end to light. Furthermore, I was always under time pressure—having to

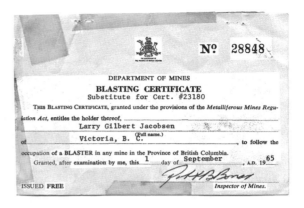

Blasting Certificate (Metalliferous Mines). The Department of Mines blasting certificate was supposedly granted upon proof that the person to whom it was issued had been working with explosives under the supervision of a qualified person for at least six months. In actual practice however, the Mine Inspector granting the certificate took the applicant's word for it and asked a few simple questions to determine the applicant's experience. One standard question was, "what is the burning speed of tape fuse?" "Forty seconds per foot" was the correct answer. Many miners, not without reason, maintained that the inspector would look in the applicant's ear, and if he did not see daylight coming through, would grant the certificate. 2006 photo.

The blasting certificate issued by the Workmen's Compensation Board (WCB) was required for all blasting in BC other than in mines. It was commonly referred to as a "surface ticket." 2007 photo.

retreat to a safe place before the first shot fired. Let's assume the fuse in the first hole to fire has been trimmed to 8.5 feet. At a burning speed of 40 seconds per foot, this leaves five minutes and 40 seconds to finish lighting, turn on the air, and climb down the ladders and exit to a safe place. The hot-wire lighter would last 90 seconds (four minutes until the first shot fired), and one was expected to be finished lighting the round by the time it burned out. I was taught to always hold a second hot-wire lighter alongside the lit one just in case I still had a fuse or two to light when the first one had burned out. A miner could not afford to have his round not finish blasting for the sake of one or two holes not fired. Not only would it reduce his earnings substantially, but his reputation, too, would suffer.

If I was a long way up in a raise, it required appreciable time to get down the ladders and over to my refuge. This added to the tension. It goes without saying that a miner had to remain cool under pressure. Under these conditions, it is not surprising if a fuse was occasionally missed by a miner with a modest amount of experience. When that happened, in addition to the round not breaking to bottom, it created the hazardous situation of unfired explosives remaining in a hole that was often hidden. Miners were sometimes killed when they drilled into and set off such explosives. I have known three men who survived such mishaps, two of whom lost their vision instead.

Thermalite Igniter Cord, which came into use in the mid-fifties where I worked, changed all that— it did away with the need to rush. Ours was a cord about 1/16 of an inch thick that burned at 16 seconds per foot. Using this product made it unnecessary to trim the fuses. The fuses could now be shorter, have an igniter attached to the end protruding from the drill hole, to which the igniter cord could easily be connected. The cord was attached to the fuses in the order in which the holes were to be fired. With this system, the miner no longer needed to light a sequence of fuses. He just had to light the igniter cord, which in turn lit each fuse attached to it. Furthermore, he could increase the length of the cord to allow himself plenty of time to reach safety.

Regardless of the method used, a miner always waited close by to count the shots as they went off. If he counted fewer shots than expected, he knew there was a problem and would of course report it to the shift boss when he came out from underground. The shift boss would then leave a warning for his cross shift. In no case was a man allowed to return to the face until a minimum of 30 minutes after the blast.

THE MINE DRY

During my years as a miner, every underground mine in BC was required by law to have a "mine dry." This was a building the main purpose of which was to dry the men's underground clothing between shifts. It contained a set of common showers, sometimes lockers for the men, often a lamp room, as well as the mine foreman's office. It usually had washing machines in which the miners could wash their "diggers" (underground clothing).

The reason for a dry was that the miners' clothes were often wet when the men came off shift—either from perspiration, from working in a wet area, or from being spattered by the water used in drilling, although some miners liked to collar their holes dry. However, inhaling the dust so generated invited silicosis—especially in gold mines where much of the rock was quartz and therefore rich in silica.

In the dry, each man hung his wet clothes on a three- or four-pronged hook tied to a rope that was threaded through a pulley suspended from the high ceiling. He hoisted the clothes up into the warmer air for drying between shifts—hence the term "dry" for this building. He then secured the other end of the rope to a hook on the wall or, where applicable, to one inside his locker.

In mines where there were no lockers one might arrive for work only to find items missing from the hook. That happened to me at United Keno Hill Mines in the Yukon, where my slickers disappeared. In 1952 at the Canex–Dodger tunnels, Floyd Fleming came off shift and found someone had appropriated his new and expensive Italian shoes. He penned a note and pinned it to the wall by his hook, asking the thief to please return the shoes. Of course, he never did get them back.

The shower area was usually one big room within the dry containing a number of showerheads. Here the crew all showered together when they came off shift. It was not unusual for a man to be startled by a burst of cold water when some mischievous buddy had sneakily turned off the hot water tap. This kind of horseplay was discouraged, for it had happened that someone turned off the cold water by mistake and scalded his friend instead.

The foreman and the shift bosses used the office from which to give instructions to the miners before they headed underground. There the miners also picked up sharpened drill bits and small tools, or on occasion purchased new slickers or boots.

The lamp room contained a charging area where the miners' lamps were recharged while the miners were off shift. Here the man picked up his assigned lamp from the charging rack. He hung the battery on his miner's belt together with other small tools and placed the lamp on his hard hat. The battery provided enough power to keep his lamp lit for eight to ten hours.

SILICOSIS

Accidents were not the only hazards a miner faced. At least for those who worked in the gold mines at Sheep Creek, silicosis also took a huge toll. No one who worked there for any length of time escaped its crippling effects. The silica lodges in the lungs and gradually reduces the amount of lung tissue available for extracting oxygen from the air to regenerate the blood. The first symptoms are similar to asthma in that the person experiences a shortness of breath. Eventually a person likely dies from pneumonia-related problems induced by the silicosis.

Silica is a common mineral found especially in quartz. Since gold is usually associated with quartz, it is not surprising that gold miners are especially prone to silicosis. Base metal mines do not as a rule contain much quartz; hence men who work in them do not often contract silicosis unless they worked in gold mines as well. The Canex mines were generally not regarded as being hazardous in this respect, but there were exceptions. The tungsten was mostly found in skarn, which is

Fitness Certificate (cover). 2007 photo.

```
Form 13
         Serial No. S.28295.
    WORKMAN'S IDENTIFICATION

Name in Full  LARRY JACOBSEN
Address  1205 Johnston St. Victoria B.C.
Unemployment Insurance No.  702-905-431
Country of Birth  Denmark
Nationality  Canadian
Married ✓ Single___ Widower___ Separated___ Divorced___

Height 5 ft. 9 in.  Date of birth Dec 9/1928
Racial colour WHITE  Colour of eyes HAZEL
Complexion MEDIUM  Colour of hair MID BROWN
Distinguishing marks and characteristics
    1" Scar (R) chin

Date Sept 10 19 65
Signature of Workman. X L Jacob

         First Certificate of Fitness
         (Silicosis and Pneumoconiosis)
         (To be filled in by W.C.B. only)
This certifies that I have examined Form 12 and Chest X-Ray Film made of
Larry JACOBSEN.
         (Full name of workman)
who is described herein and have found him fit for work in any industry where this Certificate of Fitness is required.

Date of Certification  September 10th, 1965.
Medical Referee _____ M.D.

         SUBSEQUENT CERTIFICATES
This certifies that I have examined and had Chest X-Ray Film made of the above named and have found him fit for work in any industry where this Certificate of Fitness is required.

Date of Examination_____
Place_____ B.C.
Examining Physician_____ M.D.

Date of Examination_____
Place_____ B.C.
Examining Physician_____ M.D.

N.B.—The holder of this Certificate is required to be re-examined not later than twelve months from date of first examination, and annually thereafter while engaged in any industry where this Certificate of Fitness is required.
    (Page 2)                                    (Page 3)
```

often associated with intrusive igneous (siliceous) rock. Clem Grotkowski's brother died of silicosis despite having worked at no other mine than at Canex, and having been there for only six years.

It was widely believed that no doctor in the pay of the Workmen's Compensation Board would diagnose silicosis. Rae Thomas of Ymir had worked in many mines, beginning just after he was discharged from the army. He had worked in several gold mines in Ontario, as well as at the Pioneer mine at Bralorne. In the Nelson region he had worked at the Van Roi at Silverton, the Reeves McDonald Mine at Remac, the Yankee girl at Ymir, as well as at the Emerald Mine a couple of times. Despite annual X-Rays he was never diagnosed with silicosis, until a car accident landed him in the Veteran's wing of Shaughnessy Hospital in Vancouver, where a non-WCB doctor examined him. The WCB awarded him a disability pension shortly thereafter. Another man, Al King (deceased), a former president of the Trail Local of the Mine Mill and Smelter Workers Union, fought ongoing battles with the WCB, on behalf of miners and their widows, and won many of them. He told me

Fitness Certificate (inside). A "Mines Fitness Certificate" was required for every person working in an underground mine. The Mines Act required every such person to be examined by a doctor and have a chest X-ray annually. This certificate is from my own collection (slightly reduced in size).

that any doctor who diagnosed a man with silicosis would no longer get any work from the WCB. That was a strong disincentive in an area where mining was the dominant industry.

DANGEROUS OCCURRENCES AT CANEX

VEE CUTS

In the summer of 1952, the tunnel miners were collaring the Jersey Forty-Two portal not far from the Jersey townsite. It seems that mine staff didn't give any thought to the fact that they were blasting outside, where flying rock could damage property or even kill someone. Perhaps the miners were so used to drilling a tunnel round the same old way that they always had, that no one gave it a second thought. For the crew used a "vee cut," which, as everyone knew, would throw the rock a great distance. When they fired the round, the first shots did indeed throw the rock far. Pieces of it showered the Jersey townsite, and one good-sized rock went through a roof, barely missing a woman in her kitchen.

TRAMWAY FAILURE

There was a dangerous occurrence on February 16, 1953, when the tramway cable failed near the top and allowed a number of tram buckets to run back into the lower terminal. Fortunately, everyone there was able to escape in time, and no one was injured.

ELECTRICAL BLASTING AND LIGHTNING STORMS

Blasting electrically posed its own special considerations. High voltage power lines and a number of transformers had been installed underground to provide electric current for slushers, high capacity ventilation fans, high-head water pumps, underground repair shops, locomotive battery charging stations, welders, and blasting switches. During an inspection it was noted that blasting cables ran alongside some of these power lines and were in danger of picking up induced current from them. Although the company rectified this problem, there were two incidents in 1955 when some shots at partly loaded faces were detonated spontaneously. Fortunately, no one was in the area when it happened. There had been severe lightning storms at the mine on both occasions.

After these two occurrences of premature blasts caused by lightning, the company, in consultation with senior Department of Mines' officials, developed a new set of guidelines to prevent accidental firing of blasts from lightning or any other electrical source. The most important precaution would be a policy to vacate electrical blasting areas during lightning storms. The guidelines read as follows: [16]

(1) One man specially trained and qualified should be appointed by the management to be responsible for the inspection and supervision of electric blasting methods and equipment.

(2) No connection of cap leg wires should be made until just prior to firing and then only after ascertaining that no electric storms are approaching or are in existence. The leg wires of caps in loaded holes should be short-circuited and coiled at the collar of the hole until connection of the blasting-circuit is commenced.

(3) All blasting lead lines should be insulated as well as possible and isolated as much as practicable from all possible conductors of electricity. This would include isolation from pipe-lines, air-lines, etc., as well as from power and lighting lines.

(4) Each working-place should have a separate intermediate or isolating switch.

(5) All blasting lead lines should be sectionalized as much as possible to reduce probability of induced current and potentials.

(6) "Lightning gaps" should be provided by means of removable cord jumpers with plugs and receptacles on each end.

(7)[17] The development of an automatic device for the detection and warning of both the approach and existence of electric storms would appear to be most valuable, as despite all the preceding recommended precautions it is still felt to be very important that all work in connection with electric blasting hook-ups be discontinued and the personnel removed.

Despite the forgoing, in 1969 there was yet another premature blast caused by lightning. Because of the storm, all workers in the area had been evacuated. Although no leads had been connected to the blast, one series of about 50 holes fired spontaneously.

ELECTRICAL[18]

Lightning was not the only dangerous occurrence of an electrical nature cited by the Minister of Mines. The following paragraph is quoted from an annual report.

---

16 *Quoted from the* 1955 British Columbia Minister of Mines Annual Report.
17 *According to Tony Triggs, Nick Smortchevsky did in fact invent and install such a device at the Emerald mine.*
18 *Quoted from the* 1958 British Columbia Minister of Mines Annual Report.

On June 7, 1958, the electrical foreman detailed two electricians to disconnect an oil switch in Dodger Forty Two substation 420 while he and two electricians went to substation 47-J-42 to cut off the power and disconnect a transformer. When the transformer was disconnected, the electricians informed the foreman that he could close the switch to return power to the line. This he did, forgetting that in so doing he was energizing the feeder to substation 420. Fortunately, the cables to the oil switch in substation 420 had been disconnected and were lying on the floor with the ends touching. A severe arc occurred but no one was injured. A special rule was written by the management which prohibits working on equipment unless the person working on the equipment has locked the switch controlling the circuit to that equipment in the "open" position.

MYSTERY
On January 22nd, 1957, a rubber-lined sand-pump in the lead-zinc concentrator of Canadian Exploration Limited flew apart with a loud report. This pump handles sands from the classifier next to the rod mill and pumps them to the ball mill. No one was injured, although the force of the explosion moved the 500-pound pump a distance of 4 feet. The cause of the explosion could not be determined.

ROCK FALLS
As can be seen in Table IV.3, falls of ground (rock) were an important factor in underground safety and in 1965 accounted for 18% of all underground accidents. On May 5, 1961, there was an event that could easily have been a disaster: a wedge-shaped rock slab weighing about 400 tons peeled off the back of a stope and crashed to the ground. Two miners were in the area, but were fortunately unhurt. (See Leigh Hazel interview).

OTHER
The events above were reported to the Minister of Mines. There were very likely others that were not. There was for example the case of a truck driver who twice backed his vehicle into an ore-pass but was unhurt. On one occasion he fell out of the driver's seat and landed on the rear wheel which arrested his fall. He would otherwise have fallen into the ore-pass and probably have been killed.

Poor fit. The operator backed too fast and went over the bumper log, and into the ore-pass. Courtesy of Al Nord.

MINE RESCUE TRAINING

The first mention of a Mine Rescue team in the Minister of Mines Annual Reports was in 1959, when a team competed in the West Kootenay competitions, as it would continue to do. The mine suffered only one lost time accident (LTA) each of the years from 1959 to 1961 and won the John T. Ryan Safety Trophy in 1960. Beginning in 1965, the mine sponsored two teams in the competitions, and in 1970 a team captained by Al Nord won the regional Mine Rescue competition and was the runner-up in the Provincial competition.

No one today would accept conditions in the mining industry that were standard a century earlier. As regulations became more stringent

Mine Rescue competition. Courtesy of Al Nord.

and mining companies took safety more seriously and implemented safety training on a regular basis, mine safety in British Columbia has continued to improve. Whereas accidents had been viewed as inevitable in earlier times, they later became more generally seen as caused by "breakdowns" in the safety programs.

There will probably always be accidents, but they are no longer seen as inevitable.

1970 West Kootenay first prize team. Photo: World of Placer, December 1970, courtesy of Al Nord. Courtesy of Al Nord.

Epilogue

A MINE FADES AWAY: SALMO REVISITED
by Matthew Allen[19]

On a bright summer's day, British Columbia's Kootenay Mountains fold away to the horizon in diminishing shades of blue. The forests crowd the lakes and hilltops with green profusion. But Terry Allen hears and sees only the ghosts of a mine that is history. Here where the forests and the grass have come to reclaim the ground, he hears the bustle of the work-day, the hum of the mine plant and the roar of the motorized mining equipment. This is Terry's last visit to the first Placer Dome mine he worked at. A week away from retirement, his duties as Environmental Coordinator with the Project Developments group in Vancouver brought him here last June to assess progress of reclamation measures started 40 years ago when the mine was still operating.

A young forest now stands where young families once lived and worked. Trees poke through deserted roads, and only the occasional cement slab remains to mark the location of the vanished concentrator and mine buildings. The contrast was particularly striking for Terry, who recalls the mine site as it was when he began as Tungsten Geologist in 1956. In those days, groups of buildings extended over a mile and a half of hillside, and were connected by a network of winding dirt roads.

Perched on a mountain shelf, the Emerald underground tungsten mine was acquired by Canadian Exploration, an early subsidiary of Placer Development, a predecessor of Placer Dome, in 1947. Exploration of the property revealed an extensive lead-zinc deposit which resulted in development of the adjacent Jersey Mine.

Lucrative contracts for the sale of tungsten, a mineral for hardening steel, to the United States Government during the Korean War provided Placer with record profits and set the stage for the Company's dramatic future growth. Salmo helped develop the expertise which led to large-scale open pit copper and molybdenum mining in British Columbia. Led by the so-called "Salmo College" of engineers, the mines produced a new generation of senior management. Placer

---

19 Matthew is the son of Terry Allen. The article was published in *World of Placer* in 1990.

gained the distinction of extracting two entirely different types of ore from interconnected underground mines while running a small open pit operation above, and pioneered the use of trackless mining methods. The company constructed an entire town near the twin mine sites on a mountaintop overlooking the town of Salmo. The only access to the mine sites was a stretch of switchback road ascending 2,000 feet steeply from the valley floor.

More than 1,000 people were employed during the height of construction, and at least 500 employees continued to live there afterwards. When Terry arrived as a young geological engineer, about a hundred panabode houses dotted the landscape. In addition, there were the usual mine buildings: a mill, concentrator, electrical and machine shops, offices, and a commissary.

The townsite included a community hall, a curling rink, and a skating rink. Enterprising employees constructed amenities such as a pool heated with water from the mine air compressors and a tennis court surfaced with iron tailings. There was a schoolhouse and a Directors' lodge.

Even back then, Placer did its best to reduce the impact of the mines and promote environmental reclamation. Terry recalls the site was cleared when the mines were closed in 1973, most of the buildings and machinery sold or relocated. The Company also carried out a comprehensive revegetation programme on the Salmo tailings ponds from the 1950s to the early 1970s.

Various types of vegetation were tested for growth in sandy conditions. Topsoil was spread on selected areas of the ponds, which were sown with experimentally proven seed, then watered and fertilized. Finally, the remainder of the ponds were seeded and fertilized from the air.

Seventeen years after closure, only a few mine openings and connecting roads remain to indicate the mines' location. What Nature has not reclaimed, local townsfolk have put to use. An attractive ranch style house has been built on the revegetated No. 1 tailings pond, just above a picturesque bend of the Salmo River. The other tailings pond has been converted into a combination racetrack and shooting range, complete with a clubhouse and grandstand.

For Terry, seeing the mine site virtually returned to nature is a source of both personal and professional satisfaction. He said: "It was always a beautiful place to live because of the view: we used to enjoy watching the larch trees on the hillside across the valley changing colour with the seasons. I'm glad the trees have come back so well; they belong here."

Glossary

adit Horizontal entrance to a mine from the surface.

Amex Brand name of a factory-mixed explosive called AN/FO.

AN/FO Acronym for Ammonium-nitrate and fuel oil mix—used as a much cheaper substitute for dynamite, but it cannot be detonated by a blasting cap alone. A special detonator or a stick of dynamite with a blasting cap in it is the common method of firing it.

back Roof of an underground workplace.

back-hole One of a top row of drill holes to be packed with explosives in a raise or drift.

ball mill Large, horizontal rotating cylinder in which crushed ore is pulverized by means of steel balls. Ore and water is fed into the cylinder at one end and the pulverized ore (a wet slurry) leaves it at its other end. If rods are used instead of balls, it is called a rod mill.

banjo See "muck stick."

blasting cap Small detonator used to set off explosives. It may be attached to a length of fuse that is lit by the miner, or it may be electrical and fired with an electric current.

blasting machine A government-approved hand operated device that produces an electric current used to detonate electric blasting caps.

bonus Net contract earnings, after deduction of the miner's (guaranteed) hourly wages and other chargeable expenses such as explosives.

bootleg Remnant of a blasted hole that has not broken to its bottom. It can sometimes contain undetonated explosives.

cage Vehicle used to transport workers or materials up or down a shaft—powered by a hoist. In an inclined shaft it travels on a set of rails. In a vertical shaft it is centered in the shaft by "guides." Sometimes it is incorrectly referred to as a skip.

cage tender A man in charge of the cage, who may also transport materials and/or equipment to the various levels for use by the crews there.

cap See "blasting cap."

calciminer Derogatory term for an incompetent miner.

chute Timbered structure with either a hand- or air-operated gate at the bottom of an ore pass that allows the trammer to load the muck into mine cars.

collar Start of a drill hole, shaft or raise.

C M & S See "Cominco."

Cominco Consolidated Mining and Smelting Company Ltd., also C M & S.

concentrator (mill) called this because it concentrates the ore by removing most of the waste rock.

contract earnings See bonus

conveyor Belt used to move crushed rock from the crushers to the concentrator.

crosscut	Drift driven to intersect a vein or ore body.
cross shift	Crew working on a different shift in the same workplace.
crusher	Large machine used to reduce the blasted ore fragments to a size suitable for the ball mill. The primary crusher is usually a "jaw" crusher, able to handle large pieces of rock, and the secondary one, a "cone" crusher, designed for much smaller rock.
development mining	Preparatory work, e.g. drifts, shafts and raises, used to prepare a block of ore (stope) for "production mining."
diamond drill	Machine using a hollow bit set with industrial diamonds. Used to drill holes for obtaining rock core samples, or in some cases, holes for blasting in a stope.
diggers	Slang for clothes a miner wears underground.
double-jacking	Drilling operation in which one miner holds and rotates a hand steel, while one or two miners hit the steel with a sledgehammer.
draw-point	Funnel-shaped entrance into the bottom of a stope from where the ore is drawn. A slusher is often used to scrape the muck from a series of draw-points into an ore pass.
drift	Horizontal tunnel.
drill doctor	Mechanic trained to repair air-powered mining equipment in an underground shop.
drill steel	Length of steel with a small diameter hole throughout its length for the passage of water or compressed air used to remove the drill cuttings. A (tungsten carbide?) drill bit will be attached to one end of it.
dry	High-ceilinged change house equipped with showers. A worker hangs his clothes on a hook attached to a rope that is threaded through a pulley near the ceiling and used to hoist his clothes up into the warmer air to dry them between shifts.
electric cap	Detonator set off by an electrical current. See also "blasting cap."
face	See "heading." The part of a heading where the actual drilling is done.
fan pipe	Rigid or collapsible duct used to blow fresh air into the mine to displace smoke and other noxious gasses.
farmer	Derogatory term for an inferior miner.
flotation	Process of concentrating the metals in the wet slurry coming from the ball mill with the aid of chemicals. The metals rise to the top as foam, which is then skimmed off as concentrate, dried, and shipped to a smelter for final extraction of the metals.
flotation cell	One in a series of upright vats in which the flotation process takes place.
footwall	Lower boundary between an ore body or a vein and the non-mineralized rock.
foreman	Second level of supervision and in charge of the shift bosses—also called mine captain.
galena	Most important lead-bearing ore
galvanometer	Small instrument used to measure the resistance of a blasting circuit in ohms.

Giraffe — Brand name of a hydraulic lift used by workers for reaching high places.

glory hole — open pit mine or workings that come to surface, often with vertical walls.

grizzly — Set of timbers or steel rails at the top of an ore pass that are spaced to prevent oversized rock from entering it.

gunnite — Also referred to as "shotcrete." Concrete applied to surfaces by spraying it on under pressure.

hand-steel — A hand held drill steel used to manually drill a hole in rock by striking it with a hammer.

hand-steeling — Act of drilling a hole with hand steel.

hanging wall — Upper boundary between an ore body or a vein and the non-mineralized rock.

hang-up — Rock in an ore pass that has wedged or consolidated to the point that it will not drop freely.

heading — General work area in a drift or raise.

hitch — Indentation into the rock made with a pick or chipping gun for the placement of a "sprag" or timber between opposing walls. The sprag is secured in place with wooden wedges.

hoist — Large electrical-powered single- or double-drum hoist used to transport men and equipment up and down a shaft.

honey box — Slang for box with latrine contents.

jackleg — Air-powered drill attached to an air-leg and used to drill small-diameter horizontal holes in drifts and stopes. The drill typically weighs from 40 to 70 pounds. The leg provides forward pressure on the drill.

jumbo — Self-powered vehicle fitted with two or more air- or hydraulic-operated drills—usually used in large tunnels or for "long-holing" in large stopes.

levels — Horizontally situated work areas in a mine—often referred to by elevations above sea level, e.g. 4,500, 4,550 and 4,600 levels.

lifter — One of a bottom row of drill holes to be packed with explosives in a raise or drift.

locomotive — Battery-, trolley-, or diesel-powered unit used to pull mine cars into and out of the mine. Also referred to as a "loki" or a "motor."

lode mine — Mine whose ore is contained in solid rock.

loose — Unstable or fractured overhead rock that must be scaled down so that it cannot fall on someone unexpectedly.

man-way — Portion of a raise or shaft isolated from the skipway or cage compartment and equipped with ladders for workers' access.

miner — Traditionally a trained man who actually drills and blasts the rock. Today it has come to mean anyone who works at or in a mine.

Molly Hogan — Term for a miner's quick splice for wire cable.

muck — Mining jargon for blasted rock.

muckers — Slang for rubber work boots.

mucking machine	Machine, usually air-operated, on rails, used to load muck into the mine cars.
muck stick	Slang for a short-handled shovel, also called a "banjo."
nipper	Slang for a miner's helper, or someone who delivers supplies to the miners.
nipping	Helping a miner.
oiler	Small steel oil container inserted into the airline—used to mix a small amount of oil into the air powering the air tools.
oilers	Slang for the waterproof outerwear used by miners while they are drilling. Also called slickers.
open pit	Excavation of an ore body from the surface down. If the pit becomes too deep to be practical, the mining may continue underground.
orebody	A large mass of mineral-bearing rock.
ore-pass	Raise with a chute at the bottom, from which the ore can be drawn and loaded into cars. See also waste-pass,
paper footage/loads.	Fictitious production reported by a miner to appear productive. This may be rampant where miners are paid in accordance with reported, versus measured production.
placer mine	Mine where the ore is found in alluvial gravels (deposited by flowing water).
portal	Surface entrance to a tunnel or drift.
powder	Slang for dynamite.
primer	Stick of dynamite with either an electric or fused blasting cap inserted into it.
production mining	Mining a block of ore (stope) prepared by "development" miners.
raise	Steep tunnel—usually over 45° and driven from the bottom up.
rawhiding	Skidding sacks of ore in an animal hide along a trail in the snow.
rock bolts	Steel bolts anchored in drill holes by expansion shells or epoxy. The outer end of the rod is threaded and a plate and nut are attached to prevent rock from loosening and falling. Wire mesh held in place by rock bolts is often used to additionally secure the rock, especially in underground hoist rooms and shops.
rod mill	See "ball mill."
round	Cycle at a heading—mucking out, setting up, drilling, and blasting.
safety fuse	Black cord containing a core of black powder that burns at a uniform rate of about 90 seconds per foot of length. It is attached to and used to fire a blasting cap.
scaling	Act of prying down loose rock that might endanger those working below it.
scaling bar	Bar usually made of aluminum tubing with inserts of one pointed end and one angled, lat chisel-like end used for scaling.
scheelite	Principal ore of tungsten. Scooptram Brand name of a rubber-tired machine with a large bucket used to transport muck from a face or a draw-point to a disposal point.

shaft	Steep or vertical tunnel driven from the top down.
shift boss, shifter	Lowest level supervisor. insures that the miners work is carried out as planned and is responsible for insuring that working conditions are safe for the miners.
shotcrete	See "gunnite."
single-jacking	Hand-steel operation by a single miner using a 4–lb. hammer that is swung with one hand while the drill steel is held and turned with the other. A strap is attached from the hammer handle to the miner's wrist to lessen fatigue. (See also double-jacking).
sinker	Air-powered drill used in sinking shafts.
skarn	Metamorphic rock often created by the intrusion of hydrothermal solutions in limestone, and hence associated with ore-carrying minerals.
skip	Properly refers to a small metal bucket powered by an air operated tugger hoist and used to hoist materials up a raise. It may incorrectly refer to a cage in a shaft, in which people ride.
skip tender	See cage tender.
skipway	Wooden slide on which a skip slides in a raise.
slickers	See "oilers."
slusher	Two- or three-drum hoist used to power a scraper to move muck along a drift from a face to an ore pass.
sprag	Short piece of small-diameter (4" to 6") round timber typically used for creating a work platform in a raise.
stope	Ore body that has been prepared for large-scale drilling and blasting.
stoper	Air-powered drill used to drive raises.
stull	Large timber used for timbering a raise or shaft, or for building a grizzly in an ore pass. It typically has a cross section of from 8"x8" up to 12"x12."
tailings pond	Large storage pond for the impoundment of the waste slurry left over from the concentrating process.
trackless mining	Mining carried out by rubber-tired equipment as opposed to rail mounted machines. No rails (tracks) are required.
tramming	Hauling blasted rock in a mine car.
trammer	Person who trams blasted rock.
trolley	Electrified overhead cable used to power a locomotive—similar to that once used by streetcars.
tugger	Single-drum hoist used to power a skip for hoisting materials up a raise.
vee cut	First holes to be blasted in tunnels (used where space allows). Two converging pairs of holes that "vee" towards the centre from opposing sides and are fired simultaneously. They create a face for subsequent holes to break to, but they tend to throw the broken rock a long distance and are therefore not used in timbered raises or shafts.

vein	A layer of ore, often quite extensive, but usually not very thick.
V-car	Side-dump ore car that is dumped by tipping the body to one side while the undercarriage remains on the rails.
waste	Rock devoid of commercial grade ore.
waste-pass	Same as an ore-pass, but used for waste rock only.
winze	Vertical or inclined shaft driven downward from a drift into an ore body.

Appendix 1

PLACER DEVELOPMENT LIMITED
A brief history of Placer's growth

Salmo: Jersey lead-zinc portal area. Western Miner Magazine, *August 1976, courtesy of Denis Hartland.*

The history of Placer Development Limited is a story of people, rather than a record of properties acquired and operations developed, though naturally it is in the field of mineral resource extraction that Placer has grown to its stature today.

There were very many people along the way who built Placer Development, all of them important though relatively few rose to places of prominence in the mining world. One who did, and who became Lieutenant Governor of British Columbia, was Charles Arthur Banks, one of the co-founders of the company. A letter elsewhere in this issue recalls Mr. Banks in 1915/16 when he was manager of the Jewel Mine, near Greenwood, BC.

In January 1926, in Vancouver, Charles Banks met William Addison Freeman, president of the Austral-Malay Tin Company. Banks then managed the BC Silver Company for the London-based Victoria Mining Syndicate. It appears that the two men had many common interests, and it was soon agreed that a company should be formed, with Banks as managing director, joined by Frank W Griffin (a well-known dredge designer of San Francisco) and Harold Peake (a dredging engineer who had an option on a placer gold property in Alaska), as directors.

JEWEL OF THE KOOTENAYS / 291

Salmo: The lead-zinc mill. Western Miner Magazine, August 1976, courtesy of Denis Hartland.

Vancouver was chosen as headquarters of the company, because Banks lived there and because Canadian tax laws at that time were favourable. Placer Development Limited was incorporated in British Columbia on 26 May 1926 with $200,000 in share capital and an extensive charter. Soon after this, offices were opened in San Francisco and in Sydney, Australia, and two more mining engineers were added to the board (Leslie Waterhouse, Australia, and Frank Short, US placer expert).

It was decided to concentrate initially on placer mining, but initial results from exploration of properties in Alaska, BC, the USA, and Colombia were disappointing. However, in 1928, Freeman acquired, from Cecil Levein, 1500 acres of leases containing alluvial gold on the Bulolo River and tributaries in New Guinea. After test drilling, further leases were acquired, Placer's capitalization was increased in 1929 to $500,000, and a prospecting company, Clutha Development Ltd, was acquired in Australia.

The advent of the business depression of the '30s caused the Australian government to cancel a promised road to the Bulolo Valley, but Charles Banks airlifted machinery components by Junkers G31 trimotor planes to a local airstrip, and the first re-assembled steel bucket dredge started to operate in 1932. By the end of the 1930s there were seven more dredges, hydro-electric power, machine shop, and facilities for more than 1000 employees at the site. Charles Banks received the American Mining and Metallurgical Society's gold medal in 1938 for pioneering air transportation in mining.

In 1935, when world conditions were depressed, Placer's net profit rose from $108,000 to $1.5-million. New ventures included Canadian Exploration Limited (Canex), formed to conduct mineral exploration in Canada; oil leases in Texas; gold reclamation in Australia; Pato Consolidated Gold Dredging and Asnazu Gold Dredging, in Colombia. Nechi Gold Dredging was added in 1936, and the Colombian operations, in an adverse equatorial climate, were successful.

The 1939–46 war set Placer back; mining and exploration in Australia, Canada, and the USA were abandoned. In 1942 the Bulolo aircraft and some facilities were destroyed, but the dredges survived, except for deterioration. Although Bulolo was operating again in 1947, there were various setbacks.

The Pato operations in Colombia are credited with keeping Placer alive during the war years. Field manager Vic Bramming and chief engineer Ben Barraza kept three of five dredges in operation.

THE EMERALD MINE

After the war it was decided to diversify, and ventures included coal in Australia, wood products at Bulolo, and oil in Texas. Most important was the 1947 acquisition of the Emerald underground tungsten mine near Salmo, southern British Columbia.

Large iron-stained gossans on Iron Mountain had been observed during prospecting in 1895. In 1908, John Waldbeser worked the Emerald lead-zinc mine (and produced 426 tons of lead ore valued at $7000). Iron Mountain Ltd operated the Emerald lead-zinc mine 1910–25, but production was suspended in 1926 because of low metal prices. In 1924/5 Arthur Lakes was appointed consulting geologist, and was joined by his brother Harold. Following further exploration in 1936 and 1938–40, Harold Lakes returned to the property in 1941. Scheelite was known to occur in samples from the property, and Harold Lakes discovered commercial quantities in an old adit at the north end of the Emerald.

In 1942 the Wartime Metals Corporation took over and operated the property as the Emerald Tungsten project, but this closed down in October 1943 after only six weeks of operation, and lay dormant until January 1947, when Canadian Exploration Ltd bought 41 mineral claims, buildings, and equipment. The tungsten mine was reopened in June 1947, and in its first six months of operation realized a profit of $178,000. In December 1948 the mill was converted to handle lead-zinc ore from the Jersey zone.

Salmo: Mucking machine operating a drift slash. Western Miner Magazine, August 1976, courtesy of Denis Hartland.

In 1951, during the Korean war, the federal government bought two blocks of ground with a view to a joint production of tungsten (Canex mined the ore at a fixed fee). In September 1952, Canex bought back the two blocks and enlarged the mill (built and equipped by the government). Also during 1952, provision was made to enlarge the lead-zinc mill from about 1000 to 2200 tons/day, the tungsten mill was enlarged from 250 to 650 tons/day, the mine was further developed and trackless equipment was introduced underground.

Production of Emerald tungsten continued until 1958, when the mine was closed, though tungsten was again produced in 1970–3 from the Invincible and East Dodger zones.

Lead-zinc was produced during the years 1960 to 1970. *[It should read 1950 to 1970.]*

Overall the Emerald operation, or the Salmo mine as it was also called, was a valuable contributor to Placer. One estimate gives round figures of $55-million for the value of tungsten produced, and some $80-million in lead-zinc.

One of the key people during the early Salmo period was John D Simpson, an Australian mining engineer who joined Placer in 1939. He worked in New Guinea and South America, and became president in 1957, a notable year in the company's growth. 'JD' became chairman of the board of Placer in 1964 and retired from that post in April 1975.

Both from the official accounts of the company and from talking to people who knew him, it appears that J D Simpson was highly regarded, for his technical and business ability it is true, but particularly for being able himself to work as part of a team and to inspire others, of all origins and occupations, to work well together. The expression 'team spirit' repeatedly crops up in conversation about the formative years at Salmo (when many Placer people got their grounding) and about current operations and future hopes.

Published in the Western Miner, *August, 1976; with permission of the* Northern Miner; *and thanks to Denis Hartland for loan of the magazine.*

Appendix 2

The Ore Crushing and Transportation System at
Canadian Exploration's Salmo, B. C. Operations

By A. D. McCutcheon*, C. M. McGowan**, and G. W. Walkey***[20]

(Annual General Meeting, Montreal, April, 1954)
(Transactions, Volume LVII, 1954, pp.398–410)

INTRODUCTION

THE SALMO PROPERTY of Canadian Exploration Limited is located 35 miles south of Nelson, B.C., just off the Nelson-Nelway highway. Full historical and geological details are covered in previous papers.[21] It is sufficient here to note the conditions which resulted in the installation of the Company's present ore crushing and transportation system.

In December, 1951, 250 tons of tungsten ore and 900 tons of lead-zinc ore were being produced daily. Exploration work carried out the same year had substantially increased probable lead-zinc and tungsten ore reserves. Metal prices were favourable, and consequently a programme of expansion was undertaken.

The lead-zinc orebodies, being relatively flat-lying and occurring near surface, were favourable for rapid development. However, with increasing production, transportation of ore to the lead-zinc mill, some 7,000 feet west and 2,000 feet below, was becoming a problem. In order to appreciate fully the nature of the problem it is necessary to review briefly the ore transportation system as it was early in 1952, when the Company was in the throes of expansion.

Lead-zinc ore was crushed at the mine portal and trucked one mile to the tramline terminal, from where it was trammed to the mill, another mile distant.

The tramline was operating at full capacity; production in excess of 900 tons per day had to be trucked to the mill, over a six-mile road of minus seven per cent grade. This arrangement was expensive (approximately

---

20 ** Superintendent Lead-Zinc Mining, ** Plant Superintendent, ***Assistant General Manager, Canadian Exploration Limited, Salmo, BC.*

21 *The Lead-Zinc and Tungsten Properties of Canadian Exploration Limited, Salmo, B.C., by J. D. Little, C. W. Ball, Q. G. Whishaw, and F. H. Mylrea, C'.LM., Trans., Vol. LVI, 1953, pp. 238-246.*

85 cents per ton for trucking and 60 cents per ton for truck-tramline combination) and lacked the storage, dependability, and flexibility required for increased production. It was apparent that major savings could be made by providing other means of ore transportation, and during the early months of 1952 a number of proposals were considered, which can be stated as follows:

(1) Re-location of the existing lead-zinc mill to a site nearer the mine.

(2) Replacement of the existing tramline by one of improved design and greater capacity.

(3) Continuation of truck haulage to the mill.

(4) Installation of a belt conveyor-gravity transfer system, combined with an underground crushing plant.

In April, 1952, the belt conveyor gravity transfer system was selected for installation, for the following reasons:

(1) It would reduce ore transportation costs by more than 75 per cent. This was a greater saving than could be realized by any alternative method considered.

(2) It would provide ample storage capacity, thus permitting the flexibility of mine and mill operation necessary for economical production.

OUTLINE OF UNDERGROUND CRUSHING PLANT AND BELT CONVEYOR— GRAVITY TRANSFER SYSTEM

TOPOGRAPHY AND GEOLOGY

Sixty per cent of the excavation was performed in monzonite and granite, the remainder in argillite. The surface is about half outcrop and half light overburden. Elevation difference is 2,000 feet in 7,000 feet, the greatest surface gradients being at either end. The central part, where No. 4 surface conveyor is located, is relatively flat.

ELEMENTS OF SYSTEM

Basically, the system consists of a series of underground and surface conveyors, and vertical ore passes, designed to transfer up to 3,600 tons ore per day from the lead-zinc and tungsten mines to the respective mills, crushing being performed en route. A unique feature of the system is the provision made for the ore crushing plant to handle both lead-zinc and tungsten ores.

Commencing at the mine end of the system and progressing downward to the lead-zinc mill, the main elements are as follows:

1.—Lead-zinc coarse-ore bin, capacity 5,000 tons
2.—Tungsten coarse-ore bin, capacity 2,500 tons
3.—Lead-zinc and tungsten underground crushing plant, capacity 250 to 300 tons per hour
4.—No. 1 underground conveyor, 775 feet, to carry tungsten and lead-zinc ore from the crushing plant to the initial transfer point on surface
5.—No. 2 surface conveyor, 657 feet, to tungsten and lead-zinc transfer point
6.—No. 2A surface conveyor, 207 feet, from tungsten and lead-zinc transfer point to tungsten mill
7.—No. 1 lead-zinc fine-ore bin, capacity 2,170 tons
8.—No. 3 underground conveyor, 897 feet, to No. 4 surface conveyor
9.—No. 4 surface conveyor, 2,600 feet, to No. 2 fine-ore bin, capacity 1,330 tons
10.—No. 5 underground conveyor, 530 feet, to No. 3 fine-ore bin, capacity 3,370 tons
11.—No. 6 underground and surface conveyor, 1,093 feet, to No. 4 fine-ore bin, capacity 4,800 tons
12.—No. 7 underground conveyor, 415 feet, to the lead-zinc fine ore bin

BASIC DESIGN CONSIDERATIONS

The excavation layout of the scheme was determined primarily by capacity requirements and ground topography. Other considerations which affected the design are as follows

(1) It was desired to place the crushers underground to provide adequate coarse-ore storage bins, to utilize one crushing plant for both lead-zinc and tungsten ores, and to permit economical construction of crushing plant.

(2) The combination of surface and underground conveyors incorporated in the scheme required a minimum of tunnel excavation.

(3) By driving a series of tunnels, several excavation and construction operations could be carried out simultaneously. Since the system was intended to reduce transportation costs by more than 75 per cent, the Company was anxious to complete the installation as rapidly as possible.

(4) Storage facilities adequate for several days' mill production were required, to permit flexibility in mine and crusher operation.

(5) Vertical ore bins were preferred, to minimize any tendency of the muck to hang up. The shape of the ore bins is based on the same consideration.

(6) It was desired to reduce vertical excavation to a minimum. In places where the vertical distance between tunnels was more than ample to provide the required storage, tunnels were driven at 11 degrees slope, which was the maximum permissible for the proposed weights and lengths of belts.

(7) Size of tunnels was set at 7 ft. by 7 ft. to provide room for 24-inch belt, with manway along one side.

PROJECT ORGANIZATION

EXCAVATION

The Tungsten Mining Division, having been assigned the excavation of the crushing plant chamber and coarse-ore bins, commenced preparatory drifting in June, 1952. Open cuts for No. 4 surface conveyor were started in July by the mechanical department. The fine ore bins and conveyor tunnels, which constitute the remainder of the system, were let to a contracting firm, which, after completing some 1,300 feet of drifting, relinquished the contract. On September 7th, this part of the project was taken over by the conveyor department of Canadian Exploration, which had been formed for this purpose. By this time, excavations were proceeding at six separate places and construction of surface conveyors No. 2 and 2A was under way. Schedules were revised and, to co-ordinate operations, all excavations, construction, and installations were placed under the direction of the General Superintendent. Excavation was placed on a three shift per day basis wherever practical, and construction was organized to commence immediately the various phases of excavation would permit.

CONSTRUCTION AND INSTALLATION

In the initial stage of the project, the staff of the mechanical and electrical departments executed all the planning, designing, estimating, and ordering of supplies and equipment.

The construction of No. 4, No. 2, and No. 2A conveyor supports was undertaken by the mechanical department in the summer of 1952. However, the construction and expansion programme being carried out at this time was so heavy that in January, 1953, it was decided to turn over the conveying and crushing installations to contractors for

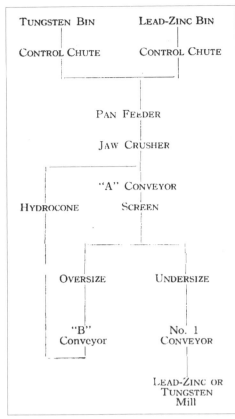

Flowchart, crushing plant.

completion. Subsequently, in accordance with Canadian Exploration specifications, the Western Knapp Engineering Company contracted to complete all crushing plant equipment, except for the pan feeder and screens. The installation of the pan feeder, screens, and all conveyor belts and hardware was contracted by the Stephens-Adamson Manufacturing Company.

The electrical department of Canadian Exploration designed and installed the power and control system for all installations.

ACCESS, AND OPERATION CONTROL

In June, 1952, two roads were constructed, one from the tungsten mine to the site of No. 3 conveyor tunnel portal and along No. 4 conveyor route, the other from the Company's south road near the lead-zinc mill to the site of No. 6 conveyor tunnel portal. These roads, which have a total length of 6,000 feet, were constructed by bulldozer, no rock work being required. A timber incline skipway 900 feet long was built to provide access from No. 6 conveyor tunnel portal to No. 5 conveyor tunnel portal. These two roads and the skipway, in conjunction with existing roads, provided access to all the tunnel portals and surface conveyor locations.

EXCAVATION

NO. 7 CONVEYOR TUNNEL

This tunnel, 7 ft. by 7 ft. in section and 380 feet long, was driven in argillite, on one per cent grade. Rounds were drilled with jack-leg machines, and mucked out using conventional track equipment. One hundred and fifty feet of tunnel, which was driven through the surface-oxidation zones, was timbered with 8-in.-square sets and 3-in. lagging.

NO. 4 FINE-ORE BIN

Feeding directly to the lead-zinc mill by No. 7 conveyor, this bin is 20 ft. by 40 ft. in cross-section by 120 ft. in height, and is designed to hold 4,800 tons of ore. From the manway raise, sub-drifts at 35-foot vertical intervals provide access to the ore bin, in the event of muck hang-up occurring during operation.

Located in argillite, the ore bin was mined by shrinkage, the adjoining manway raise being driven simultaneously. When excavation was completed, the muck was withdrawn and the walls of the ore pass were scaled and rock-bolted. About 500 one-inch-diameter rock bolts of four- and six-foot lengths were placed to prevent slabbing from the walls, which were badly fractured in places. Excavation, removal of muck, and rock bolting was completed early in February.

NO. 6 CONVEYOR TUNNEL

No. 6 tunnel, 7 ft. by 7 ft. in cross-section and 960 feet long, was driven as a scraper drift up an eleven-degree slope. Working on a three shift per day basis, each crew, which consisted of two miners and a slusher man, normally completed more than one complete cycle of drilling, mucking, and blasting. Maximum rate of advance was 120 feet per week, obtained during the final stage when the tunnel was over 900 feet long.

Two jack-leg machines were used to drill a five-foot round. A 50 h.p. electric slusher hoist, mounted 200 feet outside of the portal, was used for mucking. When the tunnel had been driven 500 feet, two scrapers, one 42-inch and one 48-inch, were used in tandem. When slushing the full 960-foot length of the tunnel, the scrapers were placed 450 feet apart. Surplus tramline cable, ¾ inch, 6-19 strand, was used with satisfactory results. Five-eighth inch cable had been tried but broke repeatedly under the heavy load. Wide-throat sheave blocks, suspended from the back of the tunnel at 100-foot intervals, were used to keep the return tail rope clear of the scrapers. Movement of the scrapers was controlled by a miner stationed near the face, who signalled the slusher man on the surface by means of an electric buzzer. The slusher man side-scraped the muck away from the portal and did the necessary nipping during the drill cycle. Ventilation was supplied by a centrifugal 20 h.p. fan, which forced fresh air to the face through 15-inch-diameter fan pipe.

NO. 3 FINE-ORE BIN

No. 3 fine-ore bin, designed for rapid excavation and to hold 3,500 tons of ore with minimum chance of hang-up, was laid out as a truncated triangular prism, 260 feet high, 10 feet wide, with a 30-foot base and a 10-foot top. A 5 ft. by 5 ft. raise was driven 30 feet above the back of No. 6 tunnel to connect with the chute raise. From there it was driven as a 50-degree knuckle back raise,

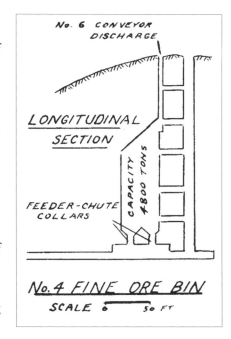

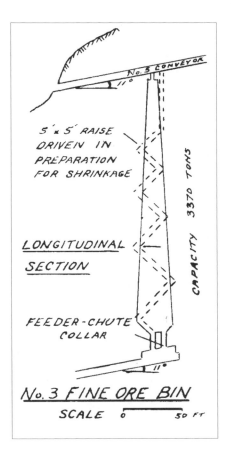

within the confines of the ore bin. This raise was driven a total length of 435 feet to break into No. 5 tunnel, 310 feet above. The bin was then excavated by shrinkage, using stopers to slash to its boundaries.

The 4,000 tons of rock excavated was removed with the same tandem scraper arrangement used in the tunnel drive. Normal mucking rate was eight tons per hour. A three-man crew was used throughout the raising and shrinking operation.

The excavation of No. 3 fine-ore bin coincided with that of No. 5 tunnel and No. 2 fine-ore bin. When the shrinkage had been completed, slashing in No. 5 tunnel was suspended for ten days while the top collar of No. 3 fine-ore bin was being concreted. A timber bulkhead was then placed on the collar, over which scraping was resumed.

INCLINED SKIPWAY FOR NO. 5 CONVEYOR ACCESS

Before No. 5 tunnel could be collared, an inclined skipway had to be constructed. Starting at No. 6 portal, this extends a distance of 900 feet up to No. 5 portal. Difference in elevation between ends is 440 feet, giving an average slope of 26 degrees.

The skipway is constructed of 4 in. by 6 in. timber bents, placed at 8-foot slope intervals, connected by 4 in. by 6 in. stringers, which support ties and 20-pound rail. Track elevation is set at a minimum of five feet above surface to allow clearance for snow. At intervals, the structure is tied by cable and, rock bolts to outcropping rock to prevent creeping.

A 50 h.p. electric 48-inch single drum hoist was installed at the lower end of the skipway. A 48-inch sheave was anchored in rock adjacent to No. 5 portal. Three-quarter-inch cable and a skip with a 4 ft. by 10 ft. deck was used. The capacity of this installation was three tons, at a hoisting speed of 250 feet per minute.

NO. 5 CONVEYOR TUNNEL

No. 5 tunnel was driven in a manner similar to No. 6, using a 50 h.p. electric slusher hoist, with tandem scraper arrangement. As the ground at the portal was inclined at 35 degrees, direct slushing was not possible. The slusher hoist was installed beside the portal, with cables pulling from sheave blocks anchored on a steel mast 20 feet high. The mast was constructed from two 8-inch channels strapped 4 inches apart, back to back. Guy cables, extended from the point of sheave anchorage to the ground below, were used to stabilize the mast. Muck built up around the base of the mast initially but, as the slope of the pile increased, it rolled progressively farther down the hill. Muck was thus easily cleared by slushing to the mast.

NO. 2 FINE-ORE BIN

The excavation programme for No. 5 tunnel and No. 2 fine-ore bin was the last to get under way, due primarily to time taken for construction of the skipway. In addition to this, 360 feet of vertical development was necessary to connect No. 5 tunnel with No. 4 tunnel, which would make this the last part of the project to be completed. These factors resulted in the design of an ore bin that could be excavated in the minimum possible time. This bin took the form of a vertical rectangular prism, 6 ft. by 14 ft. in section and 360 feet high, terminating in two cones at the bottom for chute accommodation. The top 72 feet increased to a 6 ft. by 16 ft. section to make room for No. 4 conveyor belt take-up. Raising and sinking on the bin were undertaken simultaneously.

Two 6 ft. by 6 ft. raises were collared in No. 5 tunnel, converging 20 feet above. One connected with the 6 ft. by 6 ft. manway compartment, the other with the 6 ft. by 8 ft. muck compartment of the raise. Three-inch lagging, supported by 8-inch stulls placed at 6-foot vertical intervals, separated the manway and muck compartments. The raise was driven a distance of 261 feet to meet the shaft sunk from No. 4 tunnel.

The shaft sinking installations in No.1 tunnel consisted of a 20-foot head frame, a 42-inch air-operated sinking hoist, and a 10 h.p. air slusher hoist. Muck was hoisted in a half-ton bucket and dumped in the tunnel, from where it was slushed to the portal.

Timber installed during the sinking of the shaft was designed to remain in place to accommodate the take-up of No. 4 conveyor. It consisted of standard 8-inch shaft timber sets installed down one side of the shaft, lagged with 3-inch planks to separate it from the 6 ft. by 8 ft. muck compartment.

Sinking operations were conducted on a three shift per day basis, a bench being mucked, drilled, and blasted each shift. Rate of advance was maintained at six feet per day to a depth of 90 feet, where fractured ground began to result in excessive over-break. After the timber was removed from the raise below, the final bench was blasted at a depth of 99 feet.

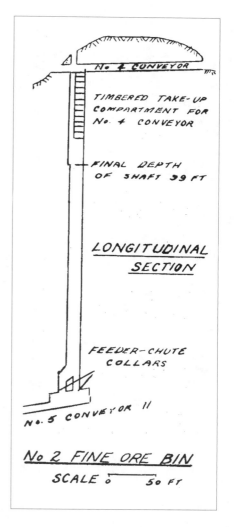

NO. 4 SURFACE CONVEYOR

No. 4 surface conveyor extends a distance of 2,600 feet across the central part of the conveyor system. With the exception of 130 feet located

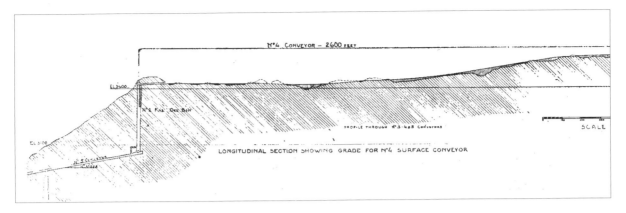

Longitudinal section showing grade for No. 4 surface conveyor.

on the west end in a 7 ft. by 7 ft. tunnel of 5 per cent grade, this conveyor conforms to a surface profile adapted for the purpose by excavating 5,000 yards of rock from open cuts, and filling where necessary. Maximum grade is 16 per cent, and minimum radius of vertical concave curves is 150 feet. Cuts and fills were made 20 feet wide, to accommodate a travel road alongside the conveyor gallery.

NO. 3 TUNNEL AND NO. 1 FINE-ORE BIN

The driving of No. 3 tunnel, 7 ft. by 7 ft. in section and 890 feet long, was performed in a manner similar to that of No. 7 tunnel. No. 1 fine-ore bin was excavated in the same manner as No. 3 fine-ore bin. Walls, which were of exceptionally stable argillite, were scaled at intervals during the muck draw, after which the timber used for a manway during this process was removed.

CRUSHING PLANT EXCAVATIONS

Crushing plant excavations performed during the period June, 1952, to February, 1953, consisted of the following operations:

(1) Driving 312 feet of tunnel, 7 ft. by 7 ft. in cross-section

(2) Slashing 800 feet of 7 ft. by 7 ft. tunnel to 9 ft. by 14 ft.

(3) Excavating gallery 40 ft. high by 74 ft. long by 14 ft. wide, for conveyors and screen

(4) Excavating crusher room 82 ft. long by 50 ft. high by 30 ft. wide.

(5) Excavating two ore pockets 30 feet in diameter, one 150 ft. high and one 96 ft. high

(6) Driving two ore pass raises a total of 813 feet

(7) Driving 373 feet of 6 ft. by 10 ft. manway raise, and 110 feet of sub-drifts to connect with ore pockets and ore pocket raises

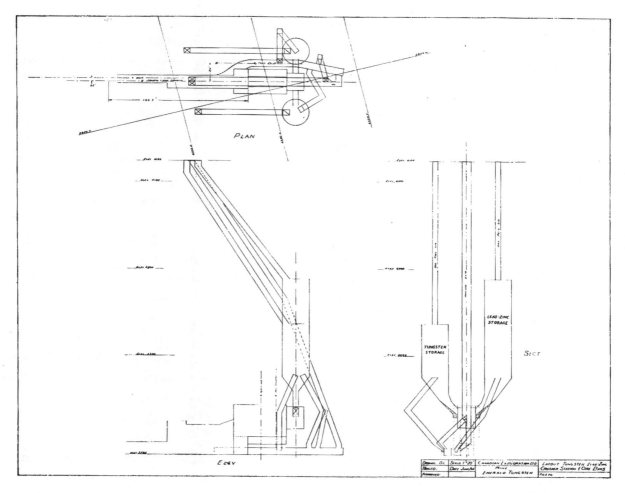

Layout, tungsten and lead-zinc crusher station and ore bins

To speed up excavation of pockets and raises, a by-pass drift was driven around the crusher room, and construction raises were driven to points just above cones of ore pockets, leaving sills below. The manway raise was driven at the same time and connected to ore pocket raises by sub-drifts. Manway and ore pocket raises were then advanced simultaneously, with access via the manway raise. The ore pockets were cut out to size by shrinkage and the muck was left in place until the crushing plant installations were complete. The sills separating the pockets from the cones were then blasted, and the muck was drawn through the chutes and fed to the crusher.

Excavation of the conveyor gallery and crusher room proceeded at the same time as the above operations. Before the removal of muck was completed, 7,000 square feet of the back and sides had been rock-bolted and gunnited. Thirteen hundred 1-inch and 1¼-inch rock bolts, in four- to eight-foot lengths, were placed. Six hundred feet of 6-inch channel, along with 4-inch square mesh screen, was bolted in place across a contact zone. Chicken wire was used over the entire area to bond the gunnite, which varied in thickness from two to five inches. Crane supports and crane were installed, muck removed, and machine and chute installations began about March 10th.

DE-WATERING NO. 2 AND NO.3 FINE-ORE BINS

During the excavation of both No. 2 and No. 3 fine-ore bins, water-flows up to 50 gallons per minute were encountered in zones of extensive jointing within the monzonite in each bin. As this amount of water would make feeder control impossible, the following action was taken.

Diamond-drill holes were drilled down from the top collar of each bin, parallel to and about 10 feet distant from the walls. Grout was pumped into these holes until pressure built up to 1,000 pounds per square inch. Approximately 10,000 feet of AX diamond-drill holes were grouted around the two ore bins in this manner. Results were a reduction in water flow in each bin to about two-thirds the original rate. The effect of the grout was most pronounced in the vicinity of the upper 100 feet of the bins and, in effect, lowered the point of ingress of water below the grouted zone.

To reduce this flow still further, drainage holes were drilled upward from the lower collar of each bin in the region just beyond the grouted zone. About 7,000 feet of EX and AX holes were drilled. Maximum flow through any drainage hole was about 90 gallons per minute. This drainage programme was terminated when the flow from each bin had been reduced to two gallons per minute. The cost of this work was in excess of $50,000.

CONSTRUCTION AND INSTALLATION

During the period July, 1952, to June, 1953, the following construction and installation was completed:

1. —7,174 feet of conveyors
2. —Three conveyor transfers
3. —2,700 feet of timber gallery for No. 4 surface conveyor and No. 6 conveyor
4. —684 feet of metal canopy for No. 2 and 2A surface conveyors
5. —Seven concrete collars for fine-ore feeders
6. —Four concrete top collars for fine-ore bins

7.—Concrete arch for ground support in No. 6 tunnel
8.—Two concrete top collars for lead-zinc and tungsten coarse-ore bin
9.—Concrete work in crushing chamber for chute collars, and machinery support and foundations
10.—Structural support for machinery and conveyors in crushing chamber.
11.—Crushing plant machinery, consisting of cone and jaw crushers, chutes, pan feeder, screens, ventilation system, and 210 feet of conveyors

Construction of footings on No. 1 surface conveyor began in July, 1952, soon after the completion of the cut-and-fill operation. By August, crews were at work on No. 2 and No. 2A conveyors. Construction of conveyor footings and feeder collars commenced in the various tunnels as soon as mining schemes world permit, beginning in No. 7 and No. 3 tunnels early in February, 1953. Construction work in the crushing chamber commenced in late February with installation of the cranes.

CRUSHING PLANT INSTALLATIONS

The underground crushing plant chamber, located in the Emerald Tungsten mine 800 feet from the 3800 level portal, is 156 feet long, has an average width of 23 feet, and a volume of 165,000 cubic feet. The plant is designed to crush, at the rate of 250 to 300 tons per hour, either lead-zinc or tungsten ore. It is an integral part of the ore transportation system, drawing mine run ore from the coarse-ore pockets and transferring the crushed product to No. 1 conveyor for removal to surface.

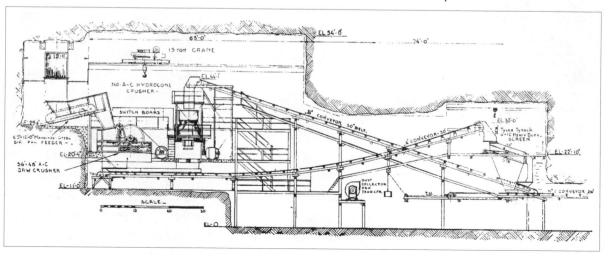

Side elevation through crushing plant.

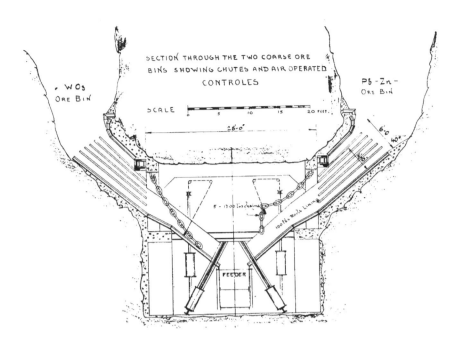

The flow-sheet of the crushing plant is shown in the accompanying diagram.

The lead-zinc and tungsten coarse-ore chutes are constructed of 3/8-inch steel plate. Each chute is mounted in a reinforced concrete collar. Concrete from the collar extends up the throat of the ore pass a distance of 15 feet. 100-lb steel rails are embedded in the concrete and extend down the sides and along the bottom of the chutes. Rate of muck flow is controlled by a set of five anchor chains, each weighing 100 pounds per foot. Muck flow can be cut off completely by an up-cutting guillotine-type chute gate. Two separate 13-inch-diameter air cylinders are used to operate the chains and gate.

A Stephens Adamson manganese-steel pail feeder, 12 ft. long by 6 ft. wide, feeds the coarse muck to the jaw crusher. This is an Allis Chalmers type A-1, 36 in. by 48 in. all steel sectionalized machine, driven by a 150 h.p., 900 r.p.m. motor. The close side setting is four inches.

The product of the jaw crusher is discharged on to "A" conveyor, which carries it to the screen. This is a Tyler 6 ft. by 12 ft. two-surface type F900 heavy duty Ty-Rock screen, the bottom deck having 7/8-inch-square openings and top deck 2-inch-square openings. The undersize discharges on to No. 1 conveyor, which is the start of the mine to mill conveyor system proper.

The oversize screen product discharges to "B" conveyor,

Stephens Adamson manganese-steel pail feeder.

Allis Chalmers hydro-cone crusher.

which feeds the secondary crusher. This is an Allis Chalmers size 7-60 hydro-cone crusher, equipped with a speed set control, reset accumulator system, and external lubricating system. The cone has a close side setting of one-half inch and has a one-inch throw. It is driven by a 200 h.p., 900 r.p.m. motor. An even distribution of feed is maintained by hanging 3/8-inch chains three inches apart around the perimeter of the conveyor discharge chute. These chains extend six inches below the wobble plate feeder of the cone.

The hydro-cone product discharges onto "A" conveyor and so operates in closed circuit with the screen.

A 15-ton single-hook electric crane is installed in the main chamber. It has a 26-foot span, 48-foot lift, and is controlled from a pendent which will reach to all floor levels.

Fresh air is drawn into the crushing chamber at the rate of 9,000 c.f.m., which is sufficient for three air changes per hour. Dust at the jaw and cone crushers, screen, and all transfer points, is collected and discharged to the portal by a 6,000 c.f.m. fan. This leaves 3,000 c.f.m. air to move along the 3800 level, so preventing mine air from entering the chamber.

CONVEYOR SYSTEM INSTALLATIONS

The total length of conveyor belt installed, 15,140 feet, is distributed as follows:

465 feet.—Conveyors A and B in the crushing chamber

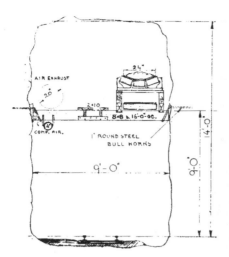

No. 1 conveyor above 3800 level, tungsten haulage tunnel.

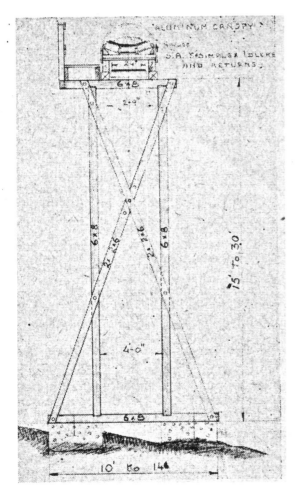

Section showing tyupe of construction of No. 2 and No. 2A conveyors.

2,920 feet.—Conveyors 1 and 2, which extend from the crushing plant to the lead-zinc and tungsten transfer point

450 feet.—Conveyor 2A, from the tungsten and lead-zinc transfer point to the tungsten mill

11,305 feet.—Conveyors 3, 4, 5, 6, and 7, which extend from the lead-zinc and tungsten transfer point to the lead-zinc mill.

No. 1 conveyor is located in the 3800 haulage adit of the Emerald Tungsten mine. It is supported on stringers, 4 in. by 8 in., laid on 8-inch-square caps which are placed at 16-foot centres, nine feet above the track. A walk-way, service pipes, and electric lines are supported alongside the conveyor. A 60-foot truss from the mine portal to the transfer house between No. 1 and No. 2 conveyors allows mine and surface vehicles to pass under the conveyor. No. 2 and No. 2A conveyors are supported by timber bents, which range in height up to 30 feet. Sectionalized aluminum covers protect the belts from the weather. Each section is four feet long, hinged on one side, and locked with spring steel clamps on the other side.

A moveable chute is installed at the lead-zinc and tungsten transfer. Depending upon the position of this chute, muck from No. 2 conveyor is dropped

Chute at lead-zinc and tungsten transfer.   *No. 4 conveyor.*

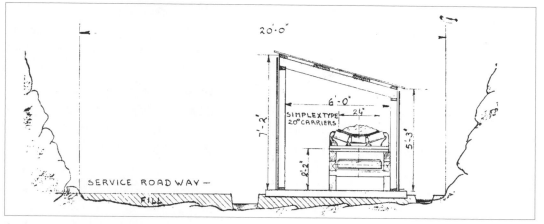

Section through No. 4 conveyor, showing gallery construction.

into No. 1 fine-ore bin, to be conveyed to the lead-zinc mill, or it is deflected onto No. 2A conveyor, which leads directly to the tungsten mill.

No. 4 conveyor extends over a distance of 2,600 feet, requiring a belt 5,270 feet long. Maximum grade is 16 per cent, and difference in elevation between ends is 145 feet in favour of the load. Maximum belt tension at the drive pulley is 4,000 pounds. The 2,500 feet of this conveyor which is on surface is enclosed in a metal-sheeted gallery, wide enough to accommodate the belt and a manway alongside. Due to change in belt length caused by varying conditions, a take-up 52 feet long is required. This is installed in a timbered compartment adjoining the upper part of No. 2 fine-ore bin.

Under full load, No. 1 conveyor is regenerative. The belt is driven by a 30 h.p., 1,800 r.p.m. motor through a Hamilton double-reduction, single helical gear speed reducer, which is connected directly to the pulley shaft.

No. 5 and No. 6 conveyors, 530 and 1093 feet long respectively,

Hamilton speed reducer, No. 4 conveyor.

Section showing construction of underground conveyor installation.

are inclined for most of their length at eleven degrees. Each belt is regenerative, and is controlled by a 75 h.p., 1800 r.p.m. motor, direct connected to a Hamilton single-helical gear reducer direct connected to the head pulley shaft.

No. 3 and No. 1 conveyors, inclined at one per cent grade, are of similar construction to No. 5 and 6 conveyors. Each conveyor, with the exception of Nos. 4, 5, and 6, has a V-belt-driven Stephens Adamson helical gear speed reducer, which is connected by chain drive to the head pulley.

All belts of the conveyor system proper are 24 inches wide. All of these except No. 4 have 3/16 inch facing and 1/16-inch backing, and are four-ply, 32-ounce duck. No. 4 conveyor has 1/8-inch facing and 1/16-inch backing, and is five-ply, 32-ounce duck. All troughing and return idlers are

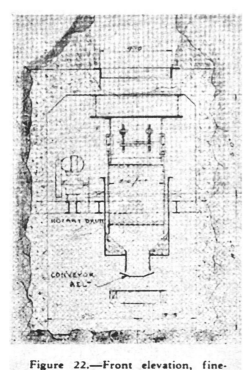
Figure 22.—Front elevation, fine-ore bin feeder installation.

Front elevation, fine-ore bin feeder installation.

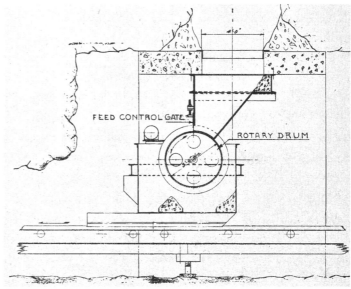
Side elevation, fine-bin feeder installation

312 / JEWEL OF THE KOOTENAYS

Fine ore feeder installed in steel hopper.

Stephens Adamson Simplex type, with 5-inch rollers equipped with Timken roller bearings. Troughing idlers are spaced at 4-foot centres, return idlers at 10-foot centres. Each conveyor has a double scraper for cleaning the belt. The scrapers have adjustable rubber wearing strips and are counterbalanced for constant pressure on the belt.

Each fine-ore feeder is of the rotary drum type, installed in a steel hopper. The hopper is suspended from the perimeter of a horizontal four-foot-square opening in the top of a concrete arch constructed over the back of the tunnel. The feed drum is 3½ feet in diameter and 3 feet wide. It is lagged with ¼-inch wear plates, made up in quarter sections to facilitate changing. It is driven at a speed of 3.58 r.p.m. by a one h.p. motor, through a Stephens Adamson reducer. The drum and hopper into which it discharges are supported by a steel frame. An adjustable gate in front of the chute hopper regulates the rate of flow of muck to the conveyor.

POWER DISTRIBUTION AND CONTROLS

Power is fed to the underground crushing plant from the surface primary 2,300-volt system, by way of a feeder. The crushers are powered by 2,300-volt wound rotor motors, with oil-filled magnetic contactors on the primary side and air-break magnetic contactors on the secondary side. Time delay relays control four-step cut-out of the starting resistors. Smaller motors are fed through a 50 k.v.a./2,300/440 volt three-phase transformer, and are controlled by across-the-line magnetic starters, located in a central Westinghouse motor control panel. All motors are push-button controlled, either from the motor control panel or from the crusher operator's platform.

The central conveyors are fed from the primary 2,300 volt system through three transformer sub-stations located at the head pulleys of conveyors 3, 4, and 5. Conveyors 1 and 2 are fed from the underground crushing plant, conveyor 2A from the tungsten mill, and conveyor 7 from the lead-zinc mill. Two vertical diamond-drill holes, having a total length of 680 feet, are used to conduct power and control cables from No. 4 to No. 5 conveyor, and from No. 5 to No. 6 conveyor. It is interesting to note that the latter hole, NX size, collared at No. 5 portal and drilled through monzonite and argillite, intersected the No. 6 tunnel 310 feet below, emerging within two feet of its surveyed destination.

For control purposes, the conveyor system is divided into two groups. The upper group, consisting of units extending from the crushing plant to the tungsten mill, has control centralized in the underground crushing plant. The other group, extending from No. 1 fine-ore bin to No. 7 conveyor, is controlled from the lead-zinc mill.

Components of either group can be run as independent units, but to facilitate operation, a system has been installed which permits any pair, or all conveyors of one group, to be interlocked. Thus, failure of one unit will automatically stop all preceding conveyors and their respective feeders. Since the crushing plant is an integral part of the transportation system, its components are interlocked with units of the conveyor system proper. Thus, the operation of the coarse-ore chute gates is controlled by the position of the tungsten and lead-zinc transfer chute, to ensure that the ore being crushed is directed to the appropriate mill. Also, the primary crusher is interlocked with the pan feeder, and the hydrocone with conveyor B to prevent overloading of the belt onto which they discharge, should it fail. This arrangement prevents unnecessary stopping of the crushers, since the amount of ore contained in them at any one time is not sufficient to overload the belt seriously.

Generally, conveyors separated by bins are run individually, and the others are interlocked. An entire group may be started by turning a special group-starting handle. Similarly, the whole group can be stopped in the proper sequence by turning an emergency stop-handle.

When the operator presses any conveyor start button, a time delay relay is energized, one set of contacts of which sets off a horn, to warn anyone near the conveyor. Another set of contacts closes 30 seconds later, bridging the starting push-button of a standard magnetic starter. The sealing of this starter is performed by a centrifugal switch, which comes into operation when the belt has reached 80 per cent of its normal speed.

By this arrangement, retarding of the belt by either slippage of the head pulley or mechanical overload will prevent the sealing of the starter and so save the belt from damage.

This belt-driven centrifugal switch closes the circuit of a master relay, which:

1.—Seals the magnetic starter in closed position
2.—Turns on a red indicating light
3.—Turns off a green indicating light
4.—Prepares the circuit of the warning buzzer in the operator's room to give a warning if the conveyor stops
5.—Introduces interlocks in preceding and following circuits to tie them together (if desired)
6.—Prepares circuit of its feeder or feeders to start.

No. 5 and No. 6 belts, which are inclined in favour of the load, are fitted with over-speed relays and thrustor brakes on their head pulleys. This is to prevent over-speeding of belts in case of electrical disturbances in supply lines, or mechanical overload on the belts.

Each belt is fitted with an ore-flow indicator which will light an amber supervisory light on the operator's control board when ore is on the belt.

A diaphragm-type switch installed in the upper part of each bin actuates an auxiliary relay when the bin is full. This relay is wired to give a visual and audible signal on the operator's board, and to stop the belt feeding into the bin.

All feeders have indicating lights, which are lighted when they are running. Feeders have separate switches to start them, but they cannot be started unless their associated conveyors are running. Unexpected stopping of feeders will also set off warning buzzers in the operator's room and extinguish their indicating lights.

All conveyors have telephones installed at their head pulleys. These are connected into the Company's automatic telephone exchange system. Lighting is installed along all conveyors and in all transfer houses. Power for welding is provided in all conveyor galleries and tunnels. Power supply for heating of exposed conveyor reduction gears is available from the same source.

METAL DETECTOR

To prevent the hydrocone crusher being damaged by tramp metal in the muck, a special detector, which responds to either magnetic or non-magnetic metal, was developed and installed by the electrical department. This device consists of two coils placed around the hydrocone feed belt, 50 feet apart. These two coils are components of a wheatstone bridge, the balance of which is upset by the passing of any metallic object of sufficient mass to damage the crusher. This produces a potential difference in the diagonal of the bridge, which after being amplified and rectified, energizes a warning device at the screen operator's station. Should a warning be received from each coil, the belt is stopped for removal of the tramp metal. The double warning prevents needless delay being caused by false warnings which may he precipitated by metallic tools in the hands of workmen.

CONSTRUCTION OPERATIONS, CONVEYOR TUNNEL FOOTINGS

Pairs of concrete pads, 12 inches square and 6 inches high, were poured on the footwall of the tunnels at intervals of 16 feet for conveyor support. Concrete was mixed at the portal of each tunnel, the mix being transported to the forms by a variety of means, according to the conditions.

In the case of No. 3 and No. 7 conveyors, track installed during the mining operations was utilized. Concrete was hauled in a ¼-yard bucket to the end of the rail from where it was distributed to nearby footings by means of wheelbarrow. As pouring proceeded, track was removed and operations retreated toward the portal.

In the case of No. 5 and No. 6 tunnels, which are inclined at 11 degrees, twenty pads were poured in sequence by bucket brigade, commencing at the portal. A section of conveyor table was then constructed over the pads when they had set sufficiently. A rubber-tired 'buggy', pulled by an air tugger hoist, was used to transport concrete to the end of the table. Additional sections were similarly constructed until the table had been extended to the end of the tunnel. An electric buzzer signal system was installed to control the movements of the buggy. The conveyor table and buggy served to transport supplies and concrete for feeder collar construction.

FEEDER COLLARS (TYPICAL INSTALLATION)

After the forms for the feeders had been placed a tugger hoist was mounted in the ore-bin cone, directly over the chute opening. Concrete was hoisted in a ¼-yard bucket from the buggy, or timber truck, and dumped into the top of the form. Movements of the bucket were controlled by buzzer signals. Total concrete required for conveyor supports, feeder, and ore-bin collars was 367 yards.

CRUSHING CHAMBER CONSTRUCTION

Muck was left in the crushing chamber following excavation to provide footing for crews placing steel and concrete for crane supports, which were located 45 feet above floor elevation. The crane was then installed and the chamber mucked out, and collars and throats of the chutes were formed. Concrete, mixed in the chamber below, was hoisted through the chutes by a tugger hoist, and dumped into the top of the forms. Concrete work for structural and equipment foundations followed. Total amount of concrete placed was 428 yards.

Installation of all equipment was completed by June 10th. Sills between chute collars and bottoms of ore pockets were blasted on June 15th, and crushers were turned over June 17th.

OPERATION

The ore crushing and transportation system is presently operating at full capacity for one shift per day, five days per week, handling 1,500 tons of lead-zinc ore and about 300 tons of tungsten ore per day. Although the system extends over a distance of 7,000 feet, by utilizing the remote controls and automatic features that have been described, the system is being operated by a crew of only six men.

This crew, consisting of a foreman, one crusher mechanic, one crusher operator, one crusher operator's helper, and two conveyor mechanics, normally operates the crushing plant and all conveyors, with the exception of No. 7 conveyor, which is controlled by the lead-zinc mill. This crew also does all normal maintenance and servicing. The crushing plant requires daily servicing and lubrication. Conveyor drives, tail pulleys, and idlers are lubricated every three months. The entire system of belts, drives, feeders, and transfers is inspected daily by the conveyor mechanics, who check for proper operation of machinery, alignment of belts, and do whatever clean-up is necessary. Muck level in the fine-ore bins is measured at the end of each shift. The measurements are telephoned to the control room of the lead-zinc mill, from where the conveyors of the lower group are operated as required to supply mill feed.

CONCLUSION

With the exception of minor difficulties encountered during the initial run-in period, operation has been satisfactory. Since the reduction in mine tonnage brought about by low metal prices, it has been possible to do all the crushing and most of the conveying in one shift per day. Thus, although the system is only handling about one-half its daily potential, very favourable costs are being realized. Crushing and conveying of lead-zinc ore during the month of January, 1954, cost 17 cents per ton (conveying 8.5 cents, crushing 8.5 cents), compared with an average cost for crushing and transportation in 1952 of 80 cents per ton. This saving is a major factor in enabling the Company to continue profitable operation during the present low metal prices.

The entire operation, which consisted of 72,000 tons excavation (including one mile of drifting, raising, and shaft sinking), construction and installation of 7,200 feet of conveyors, and the installation of a 300-ton-per-hour crushing plant, was completed in less than 13 months at a cost of approximately 1¾ million dollars.

That such a feat has been possible is due first of all to the foresight and courage of the Directors of the Company, who authorized the undertaking and guided it through to a successful conclusion. The final layout of the scheme can be attributed to a large number of persons, including both past and present staff members, who contributed in many ways to the ideas and plans that are incorporated in it. The fact that the project was completed within such a short time is a tribute to staff members who directed the operations, and to the mechanical and electrical departments, whose facilities and personnel were taxed often to the limit.

Much credit is due the Stephens-Adamson Manufacturing Company and the Western Knapp Engineering Company, who completed their operations on schedule. And finally, the part played by the miners and construction and installation men, who worked under adverse and, at times, hazardous conditions, is recognized as being, perhaps, the greatest contribution of all.

ACKNOWLEDGMENTS

The authors are indebted to Mr. N. J. Smortchevsky, Chief Electrical Engineer, who contributed notes on the power distribution and control system, and to Mr. Andre Orbeliani, Mechanical Draughtsman, who prepared most of the drawings. All photographs used in illustration were provided by the Stephens-Adamson Manufacturing Company.

Published in Canadian Mining and Metallurgical Bulletin, *Vol. 47, No. 510. (October, 1954). Reprinted with permission of the Canadian Institute of Mining, Metallurgy and Petroleum.*

Appendix 3
Placer's 25 Year Club
1976

SALMO STAFF

Clive Ball was first employed in March 1948 by Canex as a geologist at the Salmo lead-zinc and tungsten property. He transferred to Vancouver in 1954 and has worked for the exploration department continuously since then, apart from one year spent in Arizona. He is presently chief geologist.

Curly Colwell began as a warehouseman at Canex's Salmo operation in August 1951. After this practical inventory experience, he transferred to Vancouver in 1966 and is now supervisor of special projects for the purchasing department.

Jim Eastman joined Canex in November 1950 as a junior mine engineer at Salmo. He was transferred to Pato Consolidated Gold Dredging in South America in 1955 and worked there for five years until his return to Canada. Since then, he has been with the Vancouver office engineering department and is now vice-president, project development.

Doug Little joined Canex in March 1951 as a junior engineer at Salmo. He transferred to Vancouver in 1957 and held several senior positions before being appointed executive vice-president of Placer.

Herb Grutchfield is another ex-Salmo employee. He was taken on by Canex in March 1949 to work in the carpenter shop and transferred to Craigmont in 1961. He retired in 1974, at which time he was surface foreman at Craigmont.

Earl McLean joined Canex in February 1950 as an electrician at Salmo. In 1972 he transferred to the Vancouver office and is now assistant to the chief electrical superintendent in the project developments group.

Stan Hill joined Canex in August 1951 as mine shift boss at Salmo. He transferred to Craigmont in 1958, then in 1971, moved once again—this time to Gibraltar where he is now mine project foreman.

Clive Ball (right) chats with Bruce Begbie.

Gordie Donald (right) talks with Curly Colwell.

Jim Eastman and Doug Little.

Herb Crutchfield, Mora Crutchfield, Cliff Rennie, Corine McLean and Earl McLean.

Norn Steele (left) talks with Herb Grutchfield (centre) and Stan Marshall.

John Langley, Mona Crutchfield and Stan Hill.

John Langley was also first employed by Canex at the Salmo operation, where he worked in the accounting department. In 1962, he transferred to the Vancouver office and has worked in both the marketing and accounting departments. He is presently a bookkeeper with the latter.

Cliff Rennie joined Canex in September 1951 as a geologist at Salmo and transferred to Craigmont in 1957. Nine years later he moved to the Vancouver office exploration department. He spent five years with Placer Exploration in Australia from 1969 to 1974 before returning to Vancouver where he is now senior project engineer with the exploration department.

Cory Reyden worked at Salmo for 18 years after joining Canex as a mechanic in January 1947. In 1965 he moved to Endako where he is also one of the longest serving employees of that mine and now holds the position of crusher foreman.

Nick Smortchevsky joined Canex in September 1951 to work in the electrical department at Salmo. In 1961 he transferred to the Vancouver office and is now chief electrical superintendent with the project developments group.

Ross Duthie (left), Cliff Rennie (center) and Lori Fairfield (right) entertain Betty Rennie, Vivian Reyden.

NON-SALMO STAFF

Gershom Davies started his accounting career with Placer in October 1951 and has worked at the Vancouver office for 25 years. As an accountant, he has seen the accounting department grow from just a few people to its present size of about 25 employees.

Left to right: Bernice Bell, NIck Smortchevsky, Bea Marshall and Stan Marshall.

Cliff Reyen (right) chats with Ross Duthie.

Stan Marshall was hired by the Vancouver office in January 1935 as a mail boy. He spent some time in the accounting department before forming the purchasing department in 1948. As director of purchasing, he now heads a staff of about 24.

Gordie Donald has been around the Vancouver office since June 1946 when he was hired as an office junior by Placer. He has worked in the accounting department, but now handles the Company's travel desk where he arranges hundreds of travel itineraries each year.

May Donald (left) and Gershom and Anne Davies.

Bruce Begbie joined Placer's Vancouver office in January 1939 to work in the stock transfer department. (Since that time, Placer has grown so much that National Trust Company handles its stock transfer business.) Bruce has also worked in the accounting, payroll and purchasing departments, and is currently the executive driver.

Excerpted from The World of Placer — 25 Year Club, *Oct/76 (50th anniversary of Placer), courtesy of Earl McLean.*

Appendix 4

Geology of the Orebodies
By C. W. Ball*, Q. G. Whishaw,** and F. H. Mylrea[22]

GENERAL OUTLINE[23]

THE PROPERTIES are underlain by a series of sedimentary formations of late-Precambrian and lower Cambrian age which have a general strike about N.20° E. The oldest beds are quartzites and argillaceous quartzites of the Reno formation, which lie to the west of the lead-zinc orebodies. These are conformably overlain, in turn, by a series of brown argillites and limestone beds belonging to the Laib group of lower Cambrian age. The basal section of the Jersey limestone has been dolomitized and the principal lead-zinc orebodies are contained within this zone of dolomite and dolomitic limestone. The upper beds of crystalline limestone are in general barren. Succeeding the limestone on the east are beds, of black argillite, believed to be about 2,500 feet thick.

Intruding the limestones and brown argillites in the vicinity of the orebodies are tongues of granite which are offshoots from the Nelson batholith, considered to be post Jurassic in age. Adjacent to these intrusions the limestone has locally been altered to skarn rock and it is at or near such contacts that the tungsten orebodies occur.

The full thickness of the limestone is not known, but a thickness of about 800 feet is exposed on Iron Mountain. Individual beds vary greatly in thickness, due probably both to irregular deposition and to subsequent flowage.

On the basis of overfolding, it is possible to correlate the limestone which contains the Jersey lead-zinc orebodies with the limestone

---

22 *Chief Geologist, **Senior Geologist, and - §Geologist, Canadian Exploration, Limited.
23 Figures 1, 2 and 3 have been omitted. Figure 1 is a rough map of the area and can be found on page 18 of the text. Figure 2 provided an overview of the property. It is has been omitted because it is unreadable in 11" x 8½" or smaller format and Figure II.1 on page 40 of the text provides a better representation. Figure 3 provided an east-west section through the Jersey orebody, but its quality was too poor to reproduce for this book.
24 *Chief Geologist, **Senior Geologist, and - §Geologist, Canadian Exploration, Limited.

which contains the Emerald tungsten orebodies. Since marker beds are not present, the above idea must be regarded as conjectural. In support of the idea, however, it should be pointed out that large-scale overfolding of strata is one of the dominant structural features of the Salmo district.

The formations have a general strike N.20° E. and are folded along north-south axes, which plunge to the south at an average angle of 14 degrees. This plunge is reflected in the attitude of the orebodies and in the Jersey lead-zinc mine it is in places as high as 33 degrees.

The limestone and brown argillite beds are in general overfolded due to forces from the east against the underlying quartzite, which apparently acted as a resistant massif. Following this there was development of reverse faults, as for example that between the limestones and black argillites, which strikes about N.30°E., dips 40°N.W., and has a displacement probably in excess of 1,000 feet. The next stage in the tectonic history was the intrusion of bodies of granite and the attendant lead-zinc and tungsten mineralization, which was preceded and followed by normal faulting.

For a more detailed discussion of the stratigraphy and structure of the area, reference may be made to the reports by Walker (1) and Little (2, 3) cited at the end of this paper.

THE JERSEY-DODGER LEAD-ZINC ORE ZONE[25]

The Jersey lead-zinc mine is at the southern end of the property. The orebodies occur in dolomite which here forms the lower beds of the 'favourable limestone'. The dolomite has an average thickness of 140 feet and is underlain by about 10 feet of skarn rock, beneath which are brown argillites with a thickness of more than 1,000 feet. The formations have a general strike about N20°E.

The skarn rock and argillites are intensely folded, or overfolded, and crumpled, but folding in the dolomite is only moderate. The dolomite and limestone beds suffered brecciation and shattering owing to their relative incompetency, and thus afforded easy channels for the passage of mineralizing solutions. The axes of the folds plunge to the south at an average angle of 14 degrees and, in some places in the mine, as high as 33 degrees.

---

25 *By C.W. Ball and Q.G. Wishaw*

Several major faults, both reverse and normal, have been mapped on the property, and in the Jersey mine workings a number of minor normal faults and slips, some at least of which are pre-ore, have been encountered. The ore zone at the Jersey mine is bounded on the east by a strong reverse fault which follows the contact between the limestones and the overlying black carbonaceous argillites. This fault strikes N.30° E. and dips northwest at 40 degrees. The western margin of the Jersey ore zone is formed by a normal fault which parallels the granite contact. The fault strikes north-south and dips at 35° E.

An important series of normal faults strike between N.70° W. and N.45°W., with dip 50° to 60° northeast. One of these faults, however, dips southeast at 77 degrees, with the result that there is block faulting.

Other faults encountered in the mine workings strike between N.20°E. and N.30° E., with dip ranging from 60° east to vertical.

In a number of places, lamprophyre dykes follow the fault planes.

The orebodies are of the disseminated replacement type. They tend to be lenticular in cross-section, with a range in thickness from 4 feet to 60 feet (average about 12 feet), and they have considerable continuity along the strike. Up to the present, in the Jersey mine, they have been developed and mined in two open pits and in underground workings which extend for some 1,800 feet along the strike of the ore zone. Surface exploration, diamond drilling, and limited underground work, however, have shown that the ore zone continues for 7,000 feet from the Jersey mine to the Dodger claims, at the extreme northern limit of the property, and that it extends over an east-west width of 2,000 feet.

The ore minerals are galena and sphalerite. The galena, which is fine grained, is argentiferous, lead concentrates from the mill containing about 3 oz. silver per ton. The sphalerite is pale honey to amber-brown in colour, and fine grained; as indicated by assays of zinc concentrates, the mineral contains about 0.4 per cent cadmium. The average lead-zinc ratio of the ore is 1:4 and the average grade of ore milled to date has been about 2.0 per cent lead and 6.0 per cent zinc. Pyrite is present to the extent of about 5 per cent; pyrrhotite is far less abundant. The ore possesses a typical banding with rhythmic alternation of ore and dolomite.

Folding and small-scale faulting have exercised a definite structural control. Thickening of the orebodies is often noted in the synclinal troughs where overfolding is strong, and drag folding has caused abnormal thickening, which is especially pronounced in segments between faults (see Figure 4).

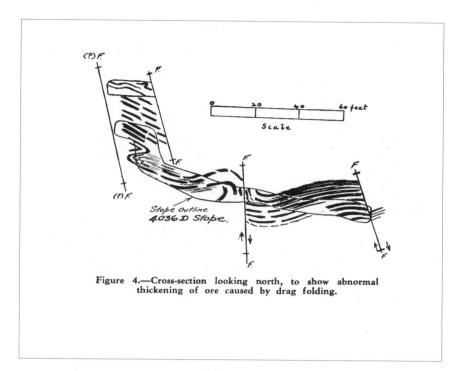

Figure 4: Cross-section looking north, to show abnormal thickening of ore caused by drag folding.

Reliable marker beds are rare in the limestone and dolomite, but the skarn serves as a very useful basal marker.

By widely-spaced surface diamond-drill holes, it has been possible to outline the general structures and the relative dimensions of the main orebodies. Current surface diamond drilling is on a grid of 400 feet by 200 feet.

Short underground diamond drill holes from the main drifts and crosscuts below the ore have given sufficient information for the laying out of stoping blocks and pillars.

Visual estimates of ore grade are made regularly and this is supplemented by regular underground sampling of fresh headings. Mine-car samples are also taken for the purpose of grade control.

Frequent inspection of the stopes and regular stope mapping have helped to keep dilution to a minimum and ensure that no ore is left unmined.

Two, glass-plate models have been prepared to show the principal diamond-drill holes and ore intersections. These are 100-scale and 50-scale. They have proved invaluable in assisting in the drilling programme and mine planning.

Present production from the mine is at the rate of 900 tons per day. From March, 1949, to August, 1952, production amounted to approximately 600,000 tons, with average grade about 2 per cent lead and 6 per cent zinc.

The initial success of the mine has been due in no small measure to the astute planning of Mr. Harold Lakes, General Manager of the Company, who directed the exploration programme.

THE EMERALD, FEENEY, AND DODGER TUNGSTEN OREBODIES[26]

GENERAL NOTE

The tungsten orebodies occur within limestone and skarn country rock and are localized where such beds conformably overlie argillite and form trough structures by contact with intrusive granite.

It is to be noted that the beds in which the tungsten ore occurs are lower, stratigraphically, than those that contain the lead-zinc deposits described in the preceding section of this paper (see Figure 8). Also that, in contrast to the lead-zinc ores, they are relatively high temperature deposits.

Underground development has shown that, for the formation of persistent ore zones, trough structures are essential. Locally, these are known as 'granite troughs' or 'tungsten troughs'. Greatest thickness of ore is found as a rule low down in the trough. However, the ore climbs up the flanks for distances up to 200 feet, and orebodies are found in subsidiary troughs on the granite flank. The limbs of tongues of granite that project upward within the troughs from the granite basement are also very favourable places for ore.

The ore occurs as scheelite in close association with pyrrhotite, biotite and white quartz. The orebodies are lenticular and the mineralization is very irregular.

At the present time, tungsten ore is produced from three mines or ore zones on the property: the Emerald, the Feeney, and the Dodger. At both the Emerald and the Dodger, the 'granite trough' with tungsten mineralization is known to extend for more than 4,000 feet along the north south strike of the formations. Up to the present, the Feeney ore zone has been tested only over a length of 250 feet.

Brief descriptions of each of these two mines follow.

THE EMERALD MINE

The Emerald tungsten deposits occur in limestone at the western mar-

---

26 By C. W. Ball and F. H. Mylrea

gin of a body of granite, elongated in a north-south direction, which here intrudes the limestone and argillites.

The typical section through the Emerald mine (Figure 5) shows how the ore occurs in limestone which conformably overlies black argillite and dips at about 50 degrees eastward into a granite contact which usually slopes to the west at a steep angle. The trough axis has an average plunge of 20 degrees southward, but the plunge and strike of the, trough are by no means regular. The southern limit of the structural pattern has not yet been determined. To date, however, it has been found to be well mineralized for a length of some 4,000 feet. Strong faulting runs along the main granite contact and minor faults, which strike sub-parallel to the trough axis, have exercised a definite structural control.

The granite surface is locally quite irregular and numerous dykes of granite and pegmatite cut across the trough. It is believed that mineralization arising from deep within the granite was distributed by the medium of quartz-tourmaline veins and siliceous veins traversing breccia zones and minor faults, and it is suggested that the strong faulting at the granite contact provided the channel for the mineralizing solutions which then found their way to structurally favourable zones within the limestone. In the vicinity of the ore zone, the granite has been largely altered to greisen and the greater part of the 'siliceous ore' in the Emerald mine is believed to be greisenic in origin.

Ore thicknesses of 30 feet are common low in the trough, with thickness diminishing up the east and west limbs. Subsidiary granite troughs on the granite flank have been highly receptive to mineralization (see Figure 6).

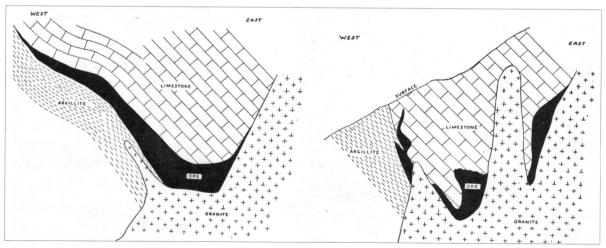

Figure 5: Cross-section of No. 1 Emerald tungsten mine, looking north.

Figure 6: Cross-section No. 2, Emerald tungsten mine, north.

The scheelite occurs in three distinct associations:

(1) Iron-Rich Ore.—This ore type consists almost entirely of pyrrhotite but invariably contains also some pyrite, quartz, and biotite. The scheelite is evenly disseminated through the ore in crystals of rather uniform size—about 1/16 inch diameter.

(2) Quartz-Rich Ore.—Quartz replacement of limestone and mineralization of the bordering greisenized granite has given rise to an ore type consisting almost entirely of quartz, with small amounts of pyrrhotite, pyrite, and sericite. The scheelite crystals are usually evenly distributed but in places they are so closely clustered that WO_3 assays of 20 per cent are obtainable over widths of one to two feet.

(3) Garnet-Rich Ore. - At the argillite-limestone contact, the beds for a width up to three feet are frequently altered to garnetite. Molybdenite is commonly present in these zones, and, as indicated by yellow fluorescence under ultra-violet ray stimulation, the scheelite is a molybdenum-bearing variety, i.e., it is an isomorphous mixture of scheelite, $CaWO_4$ and powellite, $CaMoO_4$. In this ore, the scheelite crystals are of two extreme sizes: very small and evenly disseminated, and up to an inch in diameter, the latter formed at intersections of minor fractures.

Production from the Emerald mine amounts to about 200 tons daily. Ore grade is maintained by close control of open-stope and shrinkage-stope mining. A continuous programme of detailed mapping, aided by ultra-violet light inspection, gives important data on the structural factors controlling ore localization and greatly assists the mining operation. The complex nature of the orebodies necessitates continuous development of stope faces to safeguard the production rate.

THE FEENEY MINE

The Feeney mine is approximately 600 feet north of the Emerald. The orebodies are at the north end of the same granite stock with which the Emerald ore zone is associated, but they are on its eastern side, where the granite is in contact with limestones followed to the east by argillites. The ore occurrences and structural features are similar to those at the Emerald. (Figure 7). The explored tungsten deposits have been localized in troughs formed between granite apophyses which rise from the granite basement.

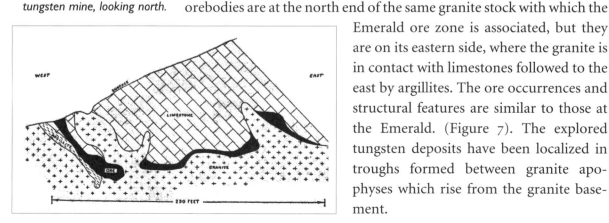

Figure 7: Cross-section, Feeney tungsten mine, looking north.

The ore zone outlined to date has a northeasterly strike and a variable plunge. It has been explored only over a length of 250 feet.

THE DODGER MINE

As stated in the *Historical Summary,* a campaign of deep diamond drilling in widely spaced holes carried out by the Company in 1950-52 on the Dodger claims at the north end of their Salmo property indicated tungsten mineralization over a length of 4,200 feet in skarn and limestone which overlie argillite and dip eastward into a steeply dipping surface of intrusive granite. In September, 1951, a drift (4,400 Main Drift) southward from the granite intersected scheelite ore at about 450 feet from the portal, and development ore was won over a length of about 400 feet along the strike of the ore zone. A month later, a cross-cut from west to east (4200 cross-cut) was started in the hillside at a point about 4,200 feet south of the portal of the Main drift. This cross-cut has since reached the projected southward continuation of the Main drift and drifting northward has commenced along the indicated ore zone (or 'granite trough'). This drift will pass beneath the ore and close to the base of the 'trough'.

Indications from surface diamond drilling are that the trough is sinuous in strike, with general trend S.15° W. The plunge is irregular and numerous sudden changes and even reversals are anticipated.

The accompanying diagrammatic longitudinal section (Figure 8) shows the relationship of the ore to the overlying limestone and dolomite, and Figure 9 shows the general structure of the trough in cross-section. As will be noted, the trough is asymmetrical, the eastern side being the steeper.

Underlying the ore on the western side is a lens of limestone overlying argillite that rests on the granite. Localization of the ore at the base of the trough appears to be controlled to a certain extent by small-scale faults and slips which are sub-parallel to the trough axis.

The orebodies are lenticular in habit, and grade is subject to rapid change. They show a tendency to split and splay-out at their margins. Thus, it is quite common to have fingers of scheelite ore separated by bands of limestone and skarn.

As exposed in present workings in the 4400 Main drift, the orebodies have an average dip of 45° east. Their average thickness is about 8 feet, with the maximum observed thickness about 35 feet.

The prevailing host-rock of the orebodies is a skarn rock containing pink garnet. Through this, white scheelite, in both fine and coarse crystals, is evenly disseminated, in close association with pyrrhotite and, in

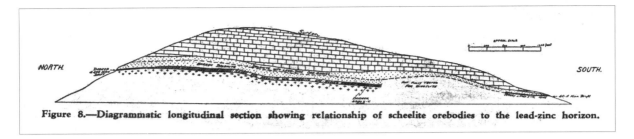

Figure 8: Diagrammatic longitudinal section showing relationship of scheelite orebodies to the lead-zinc horizon.

Figure 9: Cross-section, Dodger tungsten trough, looking north.

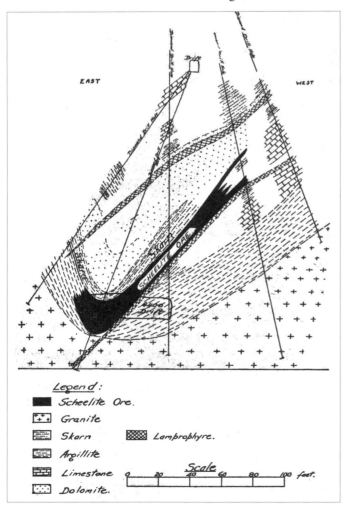

places, biotite. Quartz is common, and rarely some molybdenite is present. In places, the scheelite is a molybdenum-bearing variety.

Scheelite also occurs within the greisen underlying the main orebodies. Such mineralization is usually widely disseminated and characterized by coarse scheelite.

The mining method applied to date has been open-stoping modified by shrinkage stoping where the dip exceeds 40 degrees. Drawpoints alternate with manways at intervals of 50 feet along the line of the trough, and sub-drifts are driven on the ore from the raises.

Advance information from exploratory underground diamond-drill holes is very essential. Fanned holes are spaced at intervals of approximately 50 feet along the line of trough and the information as to structural attitudes and ore grades thus gained is vital to the mining department as a guide to planning. Three glass-plate models, showing the principal diamond-drill holes, ore intersections, and mine workings, have been used as a guide for mine planning and development work.

REFERENCES

Valuable Scheelite at the Emerald Mine discovered by Harold Lakes; Nelson Daily News, Aug. 7th, 1942, p: 9.

NORMAN, Sydney, *Most Important Strategic Find;* Spokesman Review, Spokane, Wash., Nov. 8th, 1942.

HEDLEY, M. S., *Valuable Information for Prospectors Looking for Scheelite Provided;* Financial News, Nov. 22nd, 1942.

STEVENSON, John S., *Tungsten Deposits of British Columbia;* B.C. Dept. Mines, Bull. No. 10, 1943, pp. 135146.

Emerald Tungsten Mine at Salmo is Closing: Demand for Metal Less; Nelson Daily News, Sept. 14th, 1943, p. 2.

Ottawa's Venture in Tungsten; Northern Miner, March 16th, 1944.

Emerald Tungsten; Mining World, Vol. 6, No. 7, July, 1944, pp. 15-18.

DAVIS, A. W., *$40,000 in $20 Gold Pieces Bought Emerald from Original Owners; Nelson Daily News,* Oct. 15th, 1948.

GIBSON, G., *Tungsten by the Ton;* Vancouver Sun, Apr. 12th, 1952, p. 16.

Geology of the Orebodies

(1) WALKER, J. F., *Geology and Mineral Deposits of Salmo Map-Area,_ British Columbia;* Geol. Surv. Can., Mem. 172, 1934.

(2) LITTLE, H. W., *Salmo Map-Area, British Columbia,* (Summary Account) ; Geol. Surv. Can., Paper 5019, 1950.

(3) LITTLE, H. W., *The Stratigraphy and Structure of the Salmo Map-Area, B.C.; C.I.M.,* Trans., Vol. LIV, 1951, pp. 133-136.

(4) HEDLEY, M: S., *Report on Nelson Area: Tungsten Deposits of British Columbia;* B.C. Dept. Mines; Bull No. 10, 1943, pp. 133-146.

Published in The Canadian Mining and Metallurgical Bulletin, *Vol. 46, No. 496 (Aug. 1953). Reprinted with permission of the Canadian Institute of Mining, Metallurgy and Petroleum.*

Index

A
Ackert, Joel 191, 250
Adie, L. (Joe) 30, 35, 80, 95, 105, 109, 186
Adie, Marie 35, 80, 82
adits 33, 44, 50, 272
Adolphson, Gunnar, 54, 110, 155
Adolphson, Bob 170
Ainsworth 35, 155
Alice Arm 104, 224
Alimak 50–52, 86
Allen, Terry 202, 283
Allen, Matthew 283
Amex 51, 285
AN/FO 51, 52, 285
Anderberg, Jack 66, 67, 106, 152, 154
Anderson, Brian 100
Anderson, Clarence 54
Arlington 16, 21, 68
Ash, Dick 54, 90
Ash, Grace 54
Atlas Steel Company 34

B
ball mill 56, 122, 137, 136, 255, 280, 285, 286, 288
Ball, Clive 69, 104, 105, 107, 108, 138, 319
Banks, Charles Arthur 66, 68, 83, 95, 291, 292
Barkerville 15
Batten, H.L. 38, 66
Bauman, William Douglas 255, 257, 268
beer parlour 42, 67, 154, 165
Begbie, Bruce 321
Bengert, Eddie, 216
Bennett, Premier W.A.C., 67, 209
Bing, Charlie 159
Bishop, John 4, 19, 225
Black, Gerald 141
Black Eagle 52
Blind River 109
Blondeau 173
Bluebell 15, 158, 161
Bond, Lawrence 161, 165, 191
Boyes, Doris 125
Bradley, Owen 68, 227
Bralorne 32, 136, 184, 277
Britain 15, 29, 68, 160, 161
British Columbia 5, 9, 11, 15, 19, 28, 29, 39, 184, 189, 279, 282, 283, 291–293, 331, 333
Brown Colin 52, 54, 124, 204
Brown Eileen 54
Brownies 76, 119, 124
Britannia Beach 218, 222
Bulolo River 29, 292, 293
Bunka, Terry 86
Bunka, Wally 42, 86
bunkhouse 27, 39, 44, 81, 88, 90, 91, 95, 109, 121, 148, 166, 172, 226
Burgess, Dale 234, 270
Burgess, Sonny 32, 270

C
calcites 57
Canadian Mine Services 109
Canex School of Mines 30, 49, 78, 80, 224
Carnegie 226
Cardinal River Coal 78, 79
Cariboo 15, 103, 123
Carpenter, Dr. 159, 168, 228, 259, 263
Chalmers, Beth 125
Chernoff, Alex S. 255, 258, 259
Chisan, Bob 85
Christie, G. M. (Jerry) 30, 66, 67
Clayton, Dale 100, 170
Clubine 105
CM&S 27, 28, 63, 66, 136, 285
Columbia River Mines 80
Colwell, Curly 54, 78, 198, 319
Colwell, Rhea 54, 193
commissary 41, 78, 92, 116, 124, 160, 193, 194, 209, 284
Con mine 63, 158
compressors 12, 35, 42, 75, 76, 119, 199, 284
concentrator 12, 24, 27, 31, 35, 38, 45, 46, 66, 69, 71, 122, 139, 141, 147, 280, 283–285
cone crusher 232, 286, 309, 315
Connors Diamond Drilling 181, 234, 238, 240
Consolidated Mining and Smelting Company 27, 28, 66, 136
conveyor 12, 32, 44–46, 79, 97, 113, 117, 151–153, 155, 181, 191, 208, 215, 225, 228–230, 232, 239, 285, 297–300, 303–318
conveyor tunnel 86, 232, 299, 300
Copley, Harold 123
Copper Mountain 15, 67, 105, 106, 143
Craigmont 12, 30, 49, 51, 71, 85, 109, 110, 122, 124, 126, 151, 153, 160, 188, 189, 197, 198, 203–205, 218, 319, 320
crusher 44, 45, 47, 50, 54, 97, 153, 180, 191, 229, 231–233, 247, 285, 286, 298, 304, 305, 307, 308, 309, 313–317, 320
crushing 27, 45, 46, 147, 181, 190, 230, 296, 297–300, 304, 305, 307–310, 313, 314, 316–318
Cubs 76, 124
curling 42, 76, 92, 110, 111, 115, 124, 145, 149, 158, 159, 184, 186, 209, 211, 222, 250, 284

D
D6B dozer 53
Dale Carnegie medal 235, 266
dangerous occurrences 278, 328
Dart 85, 86, 106, 154, 235, 262, 270
Davies, Gershom 321
Dawson Creek 99, 223
diamond-drill 33, 37, 106, 306, 314, 325, 330
Dickson, George 255, 256
diesel 12, 27, 33, 37, 47, 48, 50, 52, 120, 138, 147, 235, 236, 162, 287
Dingwall, Ian 215
Dixon 96, 98
Dodds 195, 204
Dodger 42 45, 47, 48, 49, 50, 106, 113, 138, 154, 155
Dodger 44 37, 45, 47, 48, 49, 61, 85, 106, 112, 122, 138, 151
Donald, Gordie 321
Douglas, Premier James 15

332 / JEWEL OF THE KOOTENAYS

Dryzmala, Frank 257
dual-purpose crusher 44
Dumptor 50, 192

E
East Dodger 31, 56, 57, 59, 294
Eastman, Jim 151, 319
Echo Bay mines 80
Ekstrom, (Swede) 85
Elliot Lake 126, 127, 139, 140
Endako 12, 30, 49, 67, 76–78, 95, 113, 114, 122, 126, 134, 204, 320
Enderburn, Lorne 243
engineering office 146, 186, 196, 198, 203, 206
England 16, 36, 68, 81, 109, 137, 149, 150, 174
Erickson, E. E. 141, 195
Erickson, Ron 141
Euclid 53, 85, 267

F
Farkas, John 246, 255, 262, 263
Feeney 38, 47, 151, 326, 328
Feeney, Bernard 123
Feeney, Joe 38
Filyk, A. 52
Fleming, Floyd 275
fine-ore bin 298, 301–304, 306, 311, 314, 317
flotation 56, 122, 286
Fort Sheppard 16, 19, 240
Frame, W. 256
Fraser Canyon 15
Fruitvale 39, 231, 249

G
Giant mine 80, 157
Gibbons, Mrs. 41
Gibraltar 12, 13, 30, 49, 71, 137, 138, 188, 264, 319
Giraffe 50, 52, 69, 111, 209, 237, 254, 255, 262, 287
Giza, Tony 153
Gold 15, 16, 21, 28, 32, 85, 94, 103, 105, 107, 123, 145, 153, 159, 180, 184, 185, 189, 190, 242, 247, 266, 267, 275–277, 291–293, 319, 331
Golac, Bob 86, 155, 243
Gold Belt 16, 85, 145, 242
Gordon, G.A. (Gerry) 43, 44, 84, 90, 91, 98, 106, 107, 122, 123, 125, 126, 127, 136, 152, 159
Gordon, Gwen 108, 159
Gorsline, Elaine 166

Gostling, Leonard Oliver 255–257
Gould, Alfred 246, 247
Gould, Billy 235
Gould, Noble 235
Government House 6, 83, 84
Graham, Phil 84
Grant, Jim and June 248
Gray, Billy 82
Gray, James (Jimmy) 111, 237
Greenwood 29, 291
Gretchen, Pat 192
Gretchen, Paul 192
Grimm, Dick 97, 98, 101
Grimwood, G.H. 31, 93
Grotkowski, Clem 207, 277
Grotkowski, Alphonse 277
Grutchfield, Herb 98, 122, 319

H
Hallbauer, Bob 70, 126, 139, 148, 184, 189, 190, 197, 202
Hallbauer, Joan 150, 184
Harold Lakes School 65, 117, 167, 248
Hartland, Colin 74
Hartland, Denis 5, 11, 69, 74, 78, 191, 295
Hartland, Lewis 71, 72, 170, 199
Haywire Nick 85
Hazel, Dale 238
Hazel, Leigh 238, 243, 280
Hazelton 65, 66
HB 216, 226, 228, 240, 243, 250
Henderson, Ken 64, 67, 91, 227
Henderson, Ann 64
Hideaway 17
Hill, Stan 319
Hinton 79
Hogarth, D.N. 160
Hope 15
Horton 109, 138
Hunter, S.J. (Stan) 29, 31, 37, 65, 67, 87

I
Idaho 35, 43, 49, 50–52
Invincible 11, 31, 48, 49, 53, 56, 57–59, 78, 108, 225, 236, 238, 240, 263, 294
Iron Mountain 11, 19, 24, 25, 31, 38, 46, 74, 123, 127, 138, 174, 293, 293, 322

J
Jedway 226
Jensen, Ole 226

Jersey townsite 79, 104, 112, 115, 123, 124, 139, 162, 204, 206, 278
Jewel 29, 291
Johannesen, Wayne Ake 237, 255, 261, 262

K
Kellogg 49, 52
Kennco 223, 224
Kettleson, Hal 136–140, 148
Kinakin 54, 99, 169, 171, 209, 225
King, Al 277
Klovance, Bonny 191
Knight, D.A. (Doug) 52, 127
Koochin 85, 162
Kootenay Bell 16, 21, 184
Kootenay Lake 15, 85, 158, 262, 268
Korean War 12, 36, 107, 137, 283, 294
Kornum, Keith 254
Kowalyshyn, A. 54, 180, 218
Kozar, John 111, 139, 143, 144

L
Lakes, Arthur 16, 25, 151, 293,
Lakes, Harold 11, 25, 30, 35, 38, 44, 65, 69, 81, 84, 106, 117, 152, 267, 148, 293, 326, 330
Langley, John 320
Larsen, Johnny 86
Larsen, Otto 204, 205
Lavin Lumber 17
Lawrence, Ed 79, 80, 83, 84, 98, 124, 126, 147, 148, 223, 267, 272
Lee, Howard 229, 231
Legg, Rollie 67, 89
Lieutenant-Governor 11, 29, 83
lightning 32, 90, 128, 182, 278, 279
Lillooet 15
Little, J.D. (Doug) 30, 102, 109, 125, 127, 137, 150, 152, 157, 198, 224, 296, 319, 323
locomotives 33, 50
London 36, 160, 161, 174, 291
Lost Creek 19, 74, 75, 97, 100, 105, 180, 182, 221
Lundeberg, Arvid (Bert) 68
Lundgren, David 242
Luscar Sterko 79

M
Magee, T.B. 30
Mahar, Merle 234
Mahonen, John 169, 209

Maike, Debora 179
main camp 37, 38, 81, 104, 109, 112, 123, 125, 167, 219
Marshall, Stan 321
Martin, Dave 42, 192
Martin, Mike 42, 43
Mason, E. E. 27
Mason, R. 63
Maxwell, H.B. (Hub) 106, 110, 124, 126, 137–139, 202
McConnell, Ross 210
McCutcheon, A. D. 47, 109, 152, 296
McDermot, Bill 125
McDiarmid, Marshall 71
McGowan, C. M. (Charlie) 126, 127, 296
McKeon, Ma 67
McLean, Carol 100, 120, 170
McLean, Earl 5, 54, 90, 96, 127, 258, 259, 319, 321
McLean, Sheila 200, 220
McNeil, B. 228, 241
Merritt 12, 30, 134, 136, 160, 182, 185, 188, 189, 195, 197, 198, 203–206, 218, 222
Mifflin, Rollie 17–19, 21
Milburn, Bill 122
Milburn, Dennis 112, 122
Milburn, Pamela 122
mine rescue 76, 88, 184, 198, 222, 281
Molly Hogan 162, 287
molybdenum 11, 12, 25, 30, 36, 39, 60, 95, 189, 283, 328, 330
Montana 52
Montgomery, Joe 23, 93
Mosses, Gil 62, 69, 77, 127, 191, 209, 218, 223, 225, 266–268, 270
Mother Lode 16, 21, 35
Murphy, Ron 66
Mylrea, F. H 296, 322, 326
N
Nelson 262, 263, 268, 277, 296, 322, 330, 331
Nelson, Sig 66, 67
batholith 110, 303, 322
Nelson–Nelway 31, 38, 184, 296
Nelway 87
New Guinea 12, 29, 30, 85, 102, 153, 179, 203, 292, 294
New York 33, 56
New Zealand 28, 84

Nickel Plate 128, 150, 151, 247
No. 2 conveyor 310
No. 2a conveyor 229, 307, 310, 311
No. 3 conveyor 300
No. 4 conveyor 215, 300, 303, 311, 312
No. 5 conveyor 300, 314
No. 6 conveyor 300, 306, 311, 312, 314
No. 7 conveyor 300, 314, 316, 317
Nord, Al 55, 79, 100, 113, 114, 169, 180, 182, 190, 215, 228, 281
Northern Miner 6, 31, 37, 41, 65, 190, 225, 331
Nugget 16, 21, 35
O
O'Connell 55, 180, 263, 266
Olson, Mel 5, 55, 87, 127, 166, 167, 169, 183
Orbeliani, Andre 130, 132
Orbeliani, Mary 131
Orbeliani, Irene 136
ore teams 21
O'Rourke, Jim 71
Owen, Dale 78, 79
Owen, Terry 220
P
Panabode 64, 65, 82, 91, 97, 101, 104, 112, 119, 121, 157, 159, 179, 185, 193, 210, 222, 284
Panagopka, John 194
Panagopka, Wally 173, 263, 266
Paradise 52, 61, 125, 180, 185, 220
Peligren, Bill 111, 143
Perry, Colonel 63, 83, 88, 109, 137, 167
Perry, Elaine 83, 167
Peters, Bill 86, 154
Peters, Hughie 203
Peters, Wes (Yutch) 86
Phipps, Ivor 29
Pillar, Chuck S. 225
pillars 50–52, 240, 244, 325
Pioneer Gold 32
Placer 5, 6, 11, 28–30, 38, 46, 57, 66, 67, 80, 88, 89, 96, 101, 110, 124, 128, 130, 134, 136, 137, 138, 153, 160, 189, 197, 203, 224, 264, 267, 283, 284, 291–295, 319–321
Pend Oreille 15–17, 194
Ponti, Louis 145–148

Postlethwaite, Anne L. 218
Powell, H. M. 30
power line 27, 35, 85, 182, 278
Precambrian 322
Q
Queen 16, 21, 85, 145
R
raise machine 52, 86
rawhiding 18, 165, 288
RD-13 Euclid truck 53
recreation 41, 42
Red Rose 66
Reeves McDonald 17, 35, 226, 277
Reid, Ben 78, 79
Remac 61, 69, 86, 111, 117, 139, 143, 144, 153, 154, 173, 195, 197, 234, 242, 247, 248, 250, 277
Rennie, Cliff 23, 102, 138, 183, 203, 320
Reno 16, 63, 68, 205, 206, 322
Reyden, Corey 55, 97, 105, 107, 320
Richie Bros. 59
Rio Tinto 223
roaster 47, 138, 165, 166
Robinson, J. W. 51, 55, 110, 126, 170, 183, 243
Rock Creek 15
Roche de Boule 66
rod mill 137, 255, 280, 285, 288
Rose, Patricia 41, 198
Rosebud Lake 192, 216
Rotter, Bob 86, 227, 235
Rotter, Frank 63, 64, 86
Rowe, Ray 55, 246
Roynon, Nelson Earle 255
Ryall, Glenda J. 218
S
Salmo Mafia 89, 124, 127, 151, 203
Salmon River 19, 31, 151
Sanderson, Betty 125
Sanford, Alice 157
Sanford, Terry 111, 157
San Francisco 29, 109, 291, 292
Schneider, Wesley 255, 261
Scooptram 52, 155, 156, 236, 246, 255, 262, 263, 288
Scouts 76, 92, 124, 212
Second Relief 16, 85
Sheep Creek 16, 19, 21, 35, 61, 85, 117, 145, 180, 181, 185, 206, 276

Shelrud, Carl 55, 218, 225, 267, 268, 270
silicosis 145, 165, 206, 248, 254, 272, 275–278
silver 15, 16, 21, 80, 93, 96, 165, 201, 223, 266, 324
Silver King 15
Simpson, J. D. 46, 106, 107, 124, 137, 138, 151, 152, 294
smelter strike 52
Smith, Enid 55, 160, 174
Smortchevsky, N. (Nick) 98, 108, 127, 128, 130, 151, 178, 279, 318, 321
Soja, George 111, 139
Sommers, Al 78
SS Minto 87
Stard, John 79, 225
Stard, Judy 79
Stard, Ron 55, 78
Stard, Tony 77
Steane, Harold A. 137, 138, 151
Steele, Phil 236, 250
Stenzel, Ernie 73, 169, 172
Stenzel, Mary 55, 79, 166
Stevens, Mildred 55
Stevens Ralph 55, 117, 169
Stringer, Bill 84
Sullivan Lake 42, 101, 221
Sultan Minerals Inc. 5, 6, 14, 51, 59–61, 126, 168, 223, 226, 227
Sutherland, George 196, 197, 204
Sutherland, Pat 204
Swedberg, Debby 218, 267
swimming pool 12, 42, 75, 77, 82, 92, 97, 110, 119, `121, 124, 127, 149, 159, 179, 193, 199, 211, 230
switchback 43, 83, 159, 172, 236, 270, 284
Sydney 29, 104, 179, 292
Sztyler, Janusz (John) 55
T
Tailings 31, 33, 42, 51, 63, 71, 151, 195, 218, 222, 225, 237, 238, 284, 289
tape fuse 272, 273
Tetz, Adin 170, 174, 190, 263, 264
Tetz, Garnet 192, 264
Tetz, Sharon 263
Tetz, Terry 190, 263
thermalite igniter cord 272, 274
Thrums 83, 144, 194
Toomey, Bill 111

townsite 14, 39, 43, 76, 79, 80, 96, 99, 104, 112, 114, 115, 123, 124, 126, 127, 139, 162, 185, 188, 200, 204, 206, 214, 215, 221, 230, 238, 278, 284
Trachesl, Heidi and Werner 127, 274, 226
Trackless 12, 33, 34, 45, 47, 49, 50, 51, 56, 58, 67, 97, 125, 126, 138, 140, 151–153, 204, 230, 262, 284, 289, 294
Trail 14, 19, 24, 28, 39, 41, 51, 52, 63, 87, 89, 100, 105, 106, 157, 158, 167, 169, 187, 188, 190, 209, 216, 217, 220, 234, 277
tramline 27, 31, 32, 38, 67, 71, 107, 108, 151, 152, 203, 296, 297, 297, 301
Triggs, Tony 83, 84, 102, 126–128, 138, 148, 186, 188, 197, 279
Triggs, Nora 84, 148, 187
Trump Giraffe 52, 254, 262
Truro, NS 234
tungsten 11, 12, 25, 27, 28, 30, 31, 33, 34–38, 44–49, 53, 56, 57, 59, 63, 66, 67, 76, 91, 93–95, 97, 104, 107–109, 114, 137–139, 147, 152, 165, 191, 197, 202, 203, 236, 276, 283, 286, 294, 296, 297, 298, 307, 308, 310, 314, 317, 323, 326, 328, 329
Tungsten mine 11, 30, 34, 36, 38, 44, 47, 49, 59, 63, 66, 91, 93, 95, 107, 124, 126, 151, 168, 172, 181, 193, 196, 218, 226, 240, 257, 293, 294, 297, 299, 300, 307, 310, 323, 326, 331
tungstic oxide 47, 49, 53, 56, 57
U
Unruh, Jim 87
uranium 34, 109
Uranium City 89
V
Vayro, Clare 121, 215
Vayro, Glenda 228
ventilation 12, 31, 32, 50, 56, 225, 272, 278, 301, 307
Verigin, Nancy 55, 193, 209
Verschoor, John 213
vertical ore bins 299
Voikin, John 250

W
Wagner Scooptram 52, 155, 156, 236, 246, 255, 262, 263, 288
Wainwright, Harold 55, 226
Wakefield, Judy 117
Waldbeser, John 11, 17, 18, 21, 37, 273
Walkey, Graham W. 125, 298
Walton, NS 191, 234, 235, 239, 243
Waneta 17, 194
War Eagle 19
Wartime Metals Corporation 11, 25, 28, 63, 66, 293
Washington State 42, 138, 149
Waterstreet, Dennis (Bucket) 225, 235, 248
WCB 166, 209, 218, 234, 266, 267, 277, 278
WCB medal 266, 267
Weber, Melba 55, 83
Weber, R.G. (Bob) 51, 52, 55, 83, 105, 126
Wegener, August 254, 255
Wilson, Bert 123, 124, 127
Wilson, Carl and Ethel 101
Wilson, Carol 101, 112, 210
Wilson, Gilbert 112, 129
Wingerak, Andy 250
winze 47, 50, 174, 290
Wishaw, Q. G 323
Wlasiuk, Steve 198
Wragge, Binky 125
Y
Yankee Girl Ymir 16, 21, 27
Yellowknife 63, 64, 80, 136, 143, 157, 158, 204
Yosky, Tony 85
Z
zinc 12, 15, 16, 21, 24, 28, 31, 33–39, 45–47, 49, 50, 52, 53, 56, 59, 63, 67, 69, 77, 91, 93, 95–97, 103, 106, 107, 138, 139, 141, 151, 152, 159, 162, 165, 173, 174, 184, 189, 190, 191, 196, 203, 226–229, 237, 239, 280, 283, 293, 294, 296–298, 300, 307, 308, 310, 311, 314, 317–319, 322–326
Zuk, Steve 110, 158–161
Zuk, Winnie 158–161
Zucco, Jack 66

ABOUT THE AUTHOR

Larry Jacobsen came to Canada as a baby in 1929, the third in a family that eventually numbered 12 children. After growing up on farms in BC and Alberta, and deciding that farming was not for him, he embarked on a nomadic career that would eventually include 46 different employers, some of which he worked for several times. He had brief spells in logging (before power saws), sawmill work and diamond drilling, before becoming an underground "hard-rock" miner for 13 years and following that up with heavy construction work.

At age 51 having used up nine lives and gone to night school for an MBA, he switched to consulting and worked for another 56 companies in construction, energy and mining.

Jacobsen now lives in Port Coquitlam and has spent his entire working life in Alberta, British Columbia and the Yukon. He retired for the last time at age 78 (from paying work). This is his third book. The first was a small book of poetry and musings that was never published. His second book was *Leaning Into The Wind: Memoirs of an Immigrant Prairie Farm Boy* published in 2004.